HOW TO INDIVIDUALIZE MATHEMATICS SUCCESSFULLY:

WITH MATERIALS FOR IMPLEMENTATION

HOW TO INDIVIDUALIZE MATHEMATICS SUCCESSFULLY: WITH MATERIALS FOR IMPLEMENTATION

DR. MILDRED L. VINSKEY

Illustrated by Diane L. Gallagher

A Hearthstone Book

Carlton Press, Inc. New York, N.Y.

Lovingly dedicated to my husband, Edward; my children, Edward, Jr., Diane, Jean and their spouses, MaryAnn, Richard and David; and my grandchildren, Jay, Jonathan and Ryan.

Manufactured in the United States of America
ISBN 0-8062-0955-0

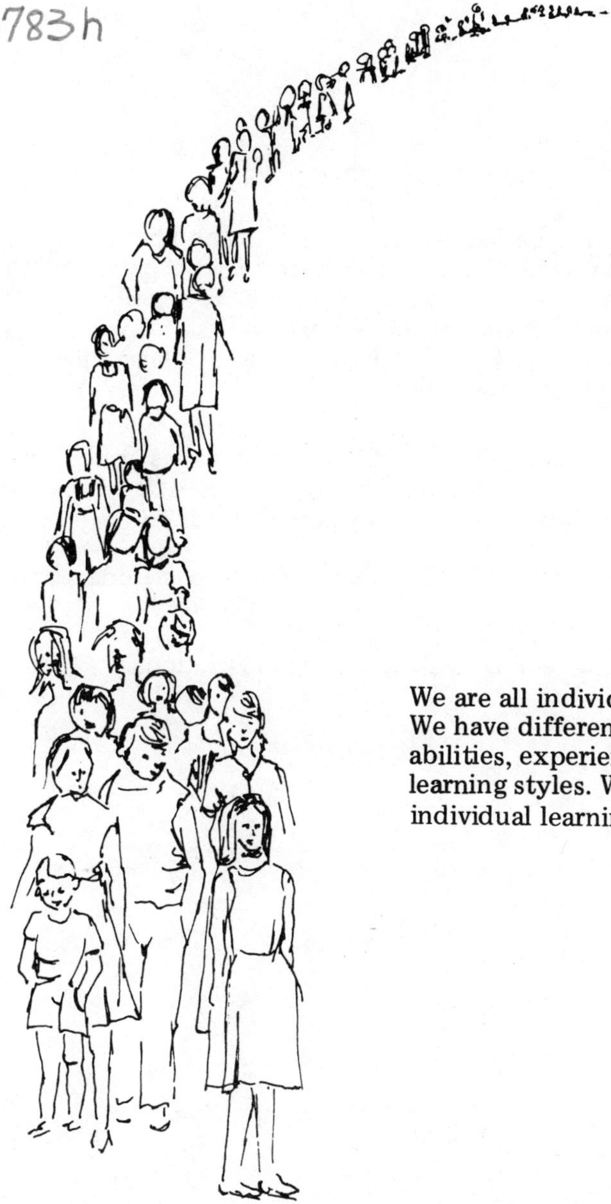

We are all individuals.
We have different backgrounds, abilities, experiences and learning styles. We all need individual learning programs!

So do I!

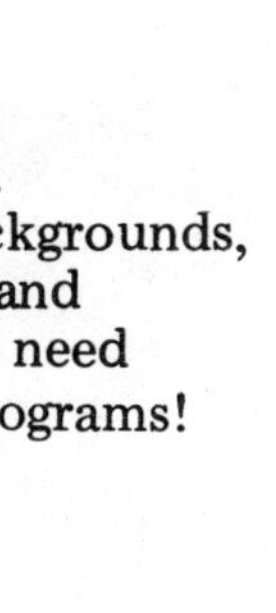

ACKNOWLEDGEMENTS

To all of my colleagues whose dedication to the philosophy of providing for individual educational needs has inspired me to compile these materials and write this book. Especially, to Dr. Francis J. Pilecki, President of Westfield State College and former Academic Dean of Fitchburg State College, for his interest and encouragement.

To the former student teachers in mathematics who have contributed to the writing of the Learning Activity Packets: Nancy Pucillo, Philip Nicastro, and Frank Guertin.

To my husband and family for their patience and understanding.

To my daughter, Diane Gallagher, for the illustrations.

CONTENTS

ESPECIALLY FOR YOU!

This book was written *ESPECIALLY FOR YOU*...

If you are a *TEACHER*, it will provide you with a ready-made, proven system for individualizing mathematics instruction in your classroom. It will relieve you of the burden of searching for hard-to-find, appropriate published materials or spending endless time and effort to create your own materials.

If you are a *STUDENT*, it will provide you with Learning Activity Packets which will meet your individual needs with relevant activities and offer you the opportunity to be selective in choosing your learning objectives.

If you are a *PARENT*, anxious to help your children with homework, it will provide you with the necessary knowledge about basic mathematics concepts, as well as introduce you to such topics as Geometry, Binary Numeration, Metric Measure, etc.

If you are *ANYONE* interested in improving your basic mathematics background, it will provide you with a variety of objectives, pretests to measure your knowledge about a particular concept, activities to reinforce your understanding, and posttests to evaluate the degree of success you have achieved.

HOW TO INDIVIDUALIZE MATHEMATICS SUCCESSFULLY

Concomitant with society's increasing emphasis upon the worth and rights of the individual is the need to provide each student with a relevant, meaningful educational program tailored to his/her wants and needs and based upon previous achievement, innate ability and learning styles.

The realization of this responsibility prompted the writer to attempt to develop and implement an effective individualized mathematics program which would maximize student-teacher contacts on a one-to-one basis and allow each student a high degree of choice and participation in the formulation of his/her own mathematics program. Coupled with this were the ever-present problems connected with the inability of students to read and understand the concepts as they are presented and explained in the mathematics textbooks, resulting in inattention, lack of interest, and frustration for both student and teacher.

Many textbook companies have developed so-called "individualized mathematics programs" which purport to provide for the needs of the individual student. However, close inspection of any of these programs usually reveals that they do offer sound behavioral objectives but then go on to suggest activities for which it is necessary to refer to pages in certain textbooks, usually those published by these very companies. Unless a teacher has these textbooks at hand, the program cannot be used. Of course, it could be argued that other textbooks could be used, but there we go!—round and round in a vicious circle, back to the textbook with each student covering the same material. The only variable that is concerned with the individual is the rate of speed at which he/she is able (or willing) to complete each activity. What happens to the student who reaches a certain point and does not progress any further because he is either unable or unwilling to make the effort? What steps are taken to diagnose why a student is unable or unmotivated to achieve a specific objective? How does the teacher find time to deal with a student whose background is so limited that he can go no further without individual help?

This is not to imply that a textbook series or multi-level materials do not have a place in the learning process, but they have limited value as a means of determining the structure of learning. What is clearly indicated here is a need for an organizational framework within which individualization can take place effectively.

The intent of this book, then, is to present a method which will help those teachers who feel frustrated in their efforts to deal with individual learning problems, who have exhausted their own creative efforts without success, or who just are not motivated to develop an individualized mathematics program of their own.

However, if you do not consider yourself to be what is generally referred to as a "dedicated" teacher, if you lack the tenacity to stick with an effort once you have started, or if you think individualization cannot work, this book is not for you.

Individualization does not just happen. It is a highly structured technique which requires a lot of time and hard work to implement and a lot more to keep it going successfully; it implies continual revision of objectives and materials; it involves planning, organization and management.

If weighed against the following outcomes, the effort will be found to be worthwhile:

1. The student will be able to experience a degree of success by working on activities commensurate with his ability and based on previous learning experiences.
2. The student will not be frustrated or haunted by fear of failure.
3. The amount of material covered will be limited only by the student's degree of motivation and rate of learning.
4. Styles of learning will be able to be accomodated through the use of filmstrips, tapes, records, etc.
5. The student will participate in choosing the objectives he will cover and the activities he will pursue to implement them.
6. All students will not need to cover the same materials at the same time.
7. Motivation will be improved, because students will be working on activities in which they are interested.
8. The relationship between students and teacher will be much closer because of frequent one-to-one discussion.
9. Evaluation of materials and student achievement will be an on-going process.

Through working with many in-service teachers in graduate courses, it has been found that all of these teachers are concerned with:

1. The inadequacy of so-called traditional methods of teaching mathematics.
2. The lack of progress of the "slow learner" and how to motivate such a student.
3. Lack of money to purchase programs and materials.
4. Lack of time to attend to individual needs.
5. Fear of administrative opposition to curriculum change.

Hopefully, the last would no longer be a concern if a teacher obtained permission to try a new program and proved that it could be carried on effectively.

THE LEARNING ACTIVITY PACKAGE (L.A.P.)

Of all the methods proposed for managing individualized instruction (small group instruction, contracts, independent study, etc.) the use of the Learning Activity Package (also called Packet) to implement objectives seems to lend itself very well to the learning of mathematics. LAPS can be very detailed, containing a rationale which explains what the packet is about, the reasons for studying it, multi-level objectives, instructions, learning experiences and self-assessment tests.

Experience has shown the writer that the simpler the LAP, the better. Therefore the following format is used in this book:

1. Statement of one or more behavioral objectives.
2. Pretest.
3. Body, containing explanations, examples, and activities.
4. Posttest.

OBJECTIVES

Since the purpose of this book is to make the individualization of mathematics easier for you, the classroom teacher, the objectives are stated and LAPS are provided to implement some objectives.

Objectives are set up for each of three levels, 6 through 8. These designations are tentative, since level 6 objectives could conceivably be successfully achieved by motivated, able 4th and 5th graders. Likewise, 9th graders who have not succeeded in mathematics and who lack fundamental knowledge in basic skill areas could very well be helped by studying any or all of these levels. In other words, the idea is to select those objectives which the student and teacher in consultation have determined would be most beneficial. Also, consideration may be given to the individual interests of the student and this is where the element of choice enters into the picture. For example, if a student is interested in geometry, he can be encouraged to practice basic computational skills through this medium. In computing perimeters of various polygons, practice can be provided in the addition of denominate numbers; solving area problems requires multiplication. There is no limit to the ways the teacher can find to accomodate the student's interest and, at the same time, review or introduce related concepts.

The choosing of objectives should not be done in a haphazard fashion. The teacher should *guide* the student and the decision should come about through agreement between student and teacher. The following points should be considered:

1. Is the student *ready* to study this particular objective? Does he/she have the necessary background to complete it successfully?

2. Does the student need to complete the objective? Many mathematics objectives fall into the area of enrichment and do not necessarily have to be completed by *all* students. For example, the study of bases other than 10 may be very difficult for some students, expecially if they are not able to understand the decimal system (base 10). It will prove frustrating to both teacher and student, if the student is forced to try to achieve the objective.

Of course, many objectives are essential and the wise teacher will focus attention upon these, whenever possible. However, continually working at the same objectives, granted they may be considered to be essential, bores and discourages the student, causing motivation to dwindle. It may be wiser to allow the student to leave certain objectives for awhile, even though he may still not have mastered them. Offer a diversion, informing the student that he/she will have to go back to the objective at a later date. This will usually be accepted and the student will often return to the objective of his own accord, when he discovers that he cannot complete other objectives successfully without this knowledge.

The teacher will have to become adept at leading and influencing students in the choice of objectives. This will come with practice. In fact, the teacher may become quite ingenious in this area, leading students toward certain objectives without the student even realizing that he is being led. Sometimes, it gets to be kind of a game with the student holding out against certain objectives and the teacher wheedling and maneuvering until the student really believes the decision was all his own. Remember that in so doing, the teacher must have the student's best interest in mind.

You will be amazed at how much you can get a student to do, using this method. Students who may have been achieving very little will respond to this kind of guidance, because it offers them a chance to experience success and eliminates the fear of continual failure. Every student will be working to some degree, maybe not all as much as you would like, but here is where *your* creativity and ability come into the picture. You must devise interesting, relevant activities that students will want to do; you must provide materials at their level of understanding; you must be genuinely interested in the subject of mathematics, so that you can pass this interest on to your students; you must give praise where praise is due.

Remember that the student must be *taught* to work towards objectives. Insist that he understands what objective he is working on and can state it in his own words. It is not sufficient to say: "I am working on Objective 7—18." Objective 7—18 (see statement of level 7 objectives) is concerned with finding Least Common Denominators which implies that the student must know what a denominator is and why it is necessary for him to find the smallest one that can be evenly divided by all of the denominators.

Some objectives must be handled separately because of level of difficulty or number of items related to the concept; others can be handled together. When providing activities, the teacher must take this into consideration. Often, several objectives can be lumped together and the student has a feeling of greater accomplishment, because he is getting more checks on the chart (this will be discussed later under record keeping).

How the choice of objectives is handled will depend upon what works best for each individual teacher in his/her own classroom. Expertise in this area will probably evolve from trial and error methods.

RECORD KEEPING

A *systematic* method of keeping records is essential for ongoing assessment of student progress. The following system has worked out very well for the writer:

1. A manilla file folder for each student. This file contains a check-off list of objectives and all of the work completed by the student. Work is not checked off and placed in this file until every error has been corrected and the teacher has approved and date-stamped it. Teacher permission is required for access to this file and it should be done under supervision. (Experience has shown that this is the best way to avoid getting files out of order, papers mislaid, etc.) This is an important file, since it gives the clearest *total* picture of the student's efforts and achievement.
2. A second set of folders can be made of colored construction paper and decorated to the student's taste. Use two pieces of 9″ x 12″ paper or fold one piece of 12″ x 18″ paper in half. Leave one long side open and seal the other edges with masking tape. To make folders last longer, reinforce open edges with masking tape. Name, grade and division should be clearly printed on each folder for immediate identification. In this folder, the student keeps his work from day to day until an objective is completed.
3. An additional set of manilla folders containing the student's name and division is first filed alphabetically and then assigned a number. The student will soon identify with his number. The files can then be placed in numerical order at the end of each lesson and will be readily found when needed for the next class. Keep the files in boxes by division. If an Industrial Arts department will cooperate, you can have boxes made to the necessary size for this purpose. These manilla folders are used to hold the colored folders when not in use, as well as extra paper and packets. This method also provides an orderly way to get a class started if students pass from class to class. Students pick up colored folders on their way into the classroom and are ready to sit down and get to work. (Numbered manilla folders remain in order in the boxes.) When class is over, if colored folders are not taken home, they can be placed in numerical order in the manilla folders on the way out. When students find it necessary or

desirable to make a new folder, the last five minutes of the class can be reserved for this. Most students take pride in the looks of their folders and wish to renew them quite often.

KEEPING DOWN COSTS

Mention was made previously of the lack of money available to purchase programs and materials. This method of individualizing mathematics is very *inexpensive*, requiring only a good supply of duplicating paper, a duplicator, fluid, masters, time, ingenuity and creativity for the initial outlay.

If you have a class of 30 students, make 10 copies of each packet. Put 5 copies out and keep 5 copies on reserve. Since all students will not be working on the same packet at the same time, it is not necessary to have many of them. The same packets can be used over and over, if students are taught to handle them with care. This procedure also eliminates clutter and waste of time. File packets that are not being used, together with the master and extra answer keys. If policy is established for their use, packets can hold up very well for 5 or 6 years, which is more than can be said for most textbooks. Students will develop a pride in keeping them neat; they do not like to use dog-eared, written on, dirty materials. The following suggestions might help:

1. Do not write in packets. Use arithmetic paper for computations and answers, numbering pages to correspond with those in packet. If there are cross-number puzzles or other drawings, they can be easily copied with tracing paper and most students will enjoy doing this as a kind of change of pace.
2. Always put packets back in place exactly as they were found. Students can take turns being in charge of the materials, seeing that they are orderly and easily accessible when needed. Again, cardboard boxes are very good for storing packets. Put objective numbers on boxes with magic markers, making numbers large so that they can easily be found.
3. Protect packets and make them attractive by using colored paper covers. Levels of packets can be easily differentiated by color coding. The writer uses pink for level 6, gold for level 7 and blue for level 8. Write the level and the objective number on the cover in large numbers with magic marker 6-18; 7-45; 8-39, etc.

CHARTS

A large chart posted on the bulletin board will give a clear picture of each student's progress as well as act as a motivation device for achievement. Students like to follow their progress. Even if it is slow, it means something to them. It should be carefully explained that there is no cause to worry or feel badly if other students have more checks than you do. There could be a number of reasons for this: the packets you are studying may be harder or longer; you may have some basics to catch up on. The main point to emphasize is that each student should be working up to his/her ability and achieving as much as possible.

This chart should contain students' names in alphabetical order and should be marked off in small squares to correspond to objective numbers. As each objective is satisfactorily completed, it is checked off and initialed by the teacher. When a packet covers more than one objective, this can be indicated by linking the checks together with curved lines.

In order to determine when certain objectives were completed, the writer uses a different color of marker to check off objectives each marking term. For example, red indicates that the objective was completed during the first marking term; blue, the second term, etc. Papers are date-stamped when completed and filed in the student's folder, so if the exact date is desired, this source can be consulted.

Thus, it can be seen that the large chart duplicates each file and safeguards the information, should something happen to any of it. After using this method over a period of five years, the writer has discovered that students are very careful with the charts. In dealing with the students in grades 7 through 9, there has never been a willful act of destruction or mutilation. They see it as an important source of information and consult it constantly.

MARKS

You will note that throughout this writing, there has been no mention of assessment in the form of marking. The reason for this is that the writer places little emphasis upon the value of marks as incentives to achieve. Accountability, yes; but marking, no.

When a student does a pretest, it is corrected but not graded. He is not punished or made to feel inferior if he is unable to do it. The purpose of a pretest is to determine whether the student has the necessary background to complete the packet. In some instances, the student may be able to go directly from the pretest to the posttest. If the student cannot do the pretest, he studies the packet first and does it later as a sort of reinforcement or extra activity before doing the posttest. Whenever he needs help, he consults the teacher. When he thinks he is ready for the posttest, he informs the teacher. Up to this point, he has access to the answer keys. Posttest answer keys are used only by the teacher and every posttest is corrected by the teacher. If there are errors, they must be corrected. The student is encouraged to try to find out why he was wrong and to correct the errors without help. The teacher judges whether the student needs more work on a particular objective. If this is the case, the student is referred to other materials.

Nowhere along the line has the student had to face up to failure in the form of a mark. If he has made a mistake, he has been allowed to correct it and has been encouraged to continue on in his learning. He is free from pressure and knows that as long as he earnestly tries and does the best he can, he will not fail. His success is being measured in terms of what he does compared to what the teacher feels he is able to do; he is not being compared with other students. He is being treated as an individual.

Of course, marks are still part of the game. Parents want them. Even though alternative methods of reporting are prevalent, they must still be accompanied by a qualifying A, B, C, D, etc. Consequently, when it is necessary to put a letter grade on a report card, the teacher must find a way to assess the work a student has done in an individualized program. The evaluation, of necessity, will be subjective on the part of the teacher. However, frequent student-teacher contacts and on-going assessment will come to the teacher's aid and a fairer mark may be arrived at than could be thought possible.

Be this as it may, it is not the intent of the writer to argue the pros and cons of marking, but to propose that marks be played down and not be used as a weapon to force children to learn. Each teacher will have to devise a method of assessment which will be in accordance with the standards of the community.

A WORD OF CAUTION

If you have reservations about individualizing mathematics or maybe feel that it is an insurmountable task, do not jump in head first. Do it gradually, step by step. Remember that you have just bought the idea yourself, and perhaps, your students will not be so easily convinced.

The main thing to remember is not to become discouraged if your plans do not fall into place *immediately*. It takes time and much planning to prepare and organize materials, to develop assessment and management techniques and to indoctrinate students into an entirely new kind of learning atmosphere. Once the whole system is ironed out, you will find that the rewards obtained in student motivation and achievement will be well worth the effort. If properly handled, you will experience the joy of helping students to succeed in areas where they have continually experienced failure in the past. Add to this your own sense of accomplishment and self-satisfaction. It will be hard; it will be demanding; but you will love every moment of it!

STEPS TO FOLLOW

STEP I. Get your materials ready. Have enough ahead for a week or two, at least, so you will not have to worry about this aspect of the implementation. Choose the objectives you wish to introduce first, basing your selection on the interests and needs of your students. It might be well to let the students select the objectives and take it from there. The more you enlist their help, the more they will be willing to cooperate and the sooner they will get into the swing of things.

The Learning Activity Packets included in this book are intended for *your* use. They may be used in any way you wish in the classroom. Copy them, change them, add to them to suit the needs of your students. Let them serve as a model to express your own creativity. These packets have all been classroom tested, so should work out well for you, too.

STEP II. Make two manilla files for each student: one for your file cabinet; the other to be kept in numerical order by divisions in boxes, as explained previously.

STEP III. Have each student make a colored folder which will be kept in the manilla file in the box. While they are doing this, explain how it will be used.

STEP IV. Make the chart that will be permanently displayed on the bulletin board.

STEP V. Make answer keys. Keep them in notebooks. Have a notebook for each level. Identify them by making large colored numbers for each level and pasting them on the front of the note book. Reinforce holes, because they are going to get a lot of use.

Have two sets: one notebook includes answers to pretests, activities, etc.; the other has answers to posttests only. Students do not have access to the posttest keys. Keep them in a special place for your own use.

STEP VI. Determine how much time you will set aside each day (or week) to implement your individualized mathematics program. Ask yourself these questions: Is your goal to individualize completely, with each student working on a program tailored to his wants and needs and based on his innate abilities and previous learning experiences? Is your goal partial individualization for some students or for all students?

STEP VII. Do a little soul-searching. How much of yourself are you willing and able to put into the effort? Will you sacrifice some of your leisure time to give the advantages of individualization to your students? Will you write packets, correct and file papers, check off objectives when you would rather be playing tennis, jogging, etc.?

STEP VIII. Now you are ready to begin. Choose students who can work well independently or with very little help from you. Set aside a certain time each day to work with these students, for example, when the rest of the class is busy doing written work, starting homework, taking tests, etc. Students may even agree to come before or after school, if they feel they have been chosen to work on a special project. You may be surprised how contagious this attitude will become. You will soon find other students begging to do work on their own. It will become kind of a status symbol, but one that will be possible for all to attain.

STEP IX. When a student has satisfactorily completed an objective, date-stamp it, initial it, check it off on the chart and file it in the cabinet.

STEP X. Discuss with the student the objective that he will study next. Help him/her to find the packet and get started.
STEP XI. Set up some rules to follow in the event that too many students need help at the same time:

1. No more than two students at the teacher's desk at a time.
2. Designate other students who can help if the teacher is too busy.

Although this may seem to be a very complicated operation as you are reading about it, you will find that you will have a smoothly running program if you follow the suggestions offered.

The writer has spent the past five years developing this program at the McKay Campus school/ Teacher Education Center, Fitchburg State College, Fitchburg, Massachusetts. Classes of 35 students have been handled on a completely individualized basis without help, with three levels of mathematics and two levels of Algebra being taught simultaneously.

So, you see, the program is a proven one. It is not just another theoretical daydream.

OBJECTIVES FOR LEVEL 6

6-1 Read and write numerals up to the trillions place.
6-2 Write powers of 10 as Base 10 numerals.
6-3 Write Base 10 numerals as powers of 10.
6-4 Write numerals in expanded notation using powers of 10.
6-5 Use the Commutative Property of Addition.
6-6 Use the Commutative Property of Multiplication.
6-7 Use the Associative Property of Addition.
6-8 Use the Associative Property of Multiplication.
6-9 Use the Distributive Property.
6-10 Write numerals in scientific notation.
6-11 Use exponents with bases other than 10.
6-12 Order operations (multiplication and division before addition and subtraction).
6-13 Add and subtract whole numbers with no limit on the number of digits or on the number of addends.
6-14 Multiply up to four digits by three digits.
6-15 Divide by 2- and 3-digit divisors with no limit on the number of digits in the dividend.
6-16 Use basic tables of linear, liquid, dry, weight, time, and temperature measure.
6-17 Change units of measure from smaller to larger units.
6-18 Change units of measure from larger to smaller units.
6-19 Add and subtract denominate numbers.
6-20 Multiply denominate numbers.
6-21 Divide denominate numbers.
6-22 Solve problems involving the four basic operations with denominate numbers.
6-23 Round off whole numbers up to the thousands place.
6-24 Use multiplication in number sentences without using the multiplication sign $(10\ (6 + 3) = N)$.
6-25 Use division in number sentences without using the division sign ($\frac{63 - 39}{4} = N$).
6-26 Use the short division method, writing answers only.
6-27 Use the "scaffold" method of long division. (Use estimated quotients and continue to subtract

until remainder is less than the divisor. Add to get final quotient and remainder, if any.)

6-28 Identify: Prime, composite, even, and odd numbers.

6-29 Use factor trees to find the prime factorization of numerals.

6-30 Use exponents to express prime factorizations.

6-31 Use exponents with bases other than 10 (2^3, 4^2, etc.).

6-32 Use ratio to compare quantities.

6-33 Identify: Common fraction, proper fraction, improper fraction, equivalent fraction, mixed numeral.

6-34 Find equivalent fractions in lower or higher terms.

6-35 Express remainders in division as fractions in lowest terms.

6-36 Find L.C.D. (Least Common Denominator) by choosing from a set of given denominators.

6-37 Change mixed numbers and whole numbers to improper fractions.

6-38 Change improper fractions to mixed numbers and whole numbers.

6-39 Add fractions with unlike denominators.

6-40 Add mixed numbers and fractions with unlike denominators.

6-41 Subtract fractions with unlike denominators.

6-42 Subtract mixed numerals with unlike denominators.

6-43 Regroup to subtract when the fraction in the minuend is smaller than the fraction in the subtrahend.

6-44 Subtract fractions from whole numbers.

6-45 Subtract fractions from mixed numbers.

6-46 Subtract mixed numbers from whole numbers.

6-47 Use symbols for "is equal to," "is less than," "is greater than" with whole numerals, mixed numerals and fractions.

6-48 Use cross-multiplication to determine whether two ratios are equivalent.

6-49 Find common factors.

6-50 Multiply a fraction by a fraction.

6-51 Multiply a fraction by a whole number.

6-52 Multiply a fraction by a mixed number.

6-53 Multiply a mixed number by a whole number.

6-54 Multiply a mixed number by a mixed number.

6-55 Find reciprocals of given numbers.

6-56 Divide a whole number by a fraction.
6-57 Divide a fraction by a fraction.
6-58 Divide a mixed number by a whole number.
6-59 Divide a mixed number by a fraction.
6-60 Divide a mixed number by a mixed number.
6-61 Solve problems involving the multiplication and division of fractions and mixed numbers.
6-62 Use the vertical method of multiplying by fractions and mixed numbers.
6-63 Add decimal fractions up to the ten-thousandths place.
6-64 Subtract decimal fractions up to the ten-thousandths place.
6-65 Solve problems involving the addition and subtraction of decimal fractions.
6-66 Multiply decimal fractions up to the ten-thousandths place.
6-67 Divide decimal fractions by whole numbers.
6-68 Divide whole numbers by decimal fractions.
6-69 Divide decimal fractions by decimal fractions.
6-70 Solve problems involving the multiplication and division of decimal fractions.
6-71 Multiply by powers of 10 by moving the decimal point.
6-72 Divide by powers of 10 by moving the decimal point.
6-73 Round off decimals in division answers to specified places.
6-74 Square a number.
6-75 Find square roots of perfect squares by estimation.
6-76 Locate integers on the number line.
6-77 Solve simple equations.
6-78 Solve simple inequalities.
6-79 Use divisibility rules (2, 3, 4, 5, 6, 8, 9, 10).
6-80 State the place value of numerals in Base 2 up to the sixth place.
6-81 Change numerals from Base 2 to Base 10.
6-82 Change numerals from Base 10 to Base 2.
6-83 Write numerals up to 50 in Base 2.
6-84 Add and subtract in Base 2.
6-85 Multiply and divide in Base 2.
6-86 Define the prefixes used in the Metric System.

6-87 Express metric measures as powers of 10.
6-88 Use linear measure in the Metric System.
6-89 Make a scale drawing.
6-90 Construct line and bar graphs.
6-91 Express ratios as parts of 100 and change to percents.
6-92 Use a proportion to find a percent of a number.
6-93 Use a proportion to find what percent one number is of another.
6-94 Use a proportion to find 100% (base) of a number when a part of the number is given.
6-95 Use proportions to solve problems involving the three types of percent.
6-96 Identify: Point, ray, line, segment.
6-97 Identify: Circle, semi-circle, radius, diameter, arc.
6-98 Use a compass.
6-99 Identify: Degree, angle, vertex, right angle, straight angle, obtuse angle, acute angle.
6-100 Use a protractor.
6-101 Measure angles of various types and sizes.
6-102 Construct angles to a specified number of degrees.
6-103 Construct the perpendicular bisector of a segment.
6-104 Determine whether two angles are congruent.
6-105 Determine whether two triangles are congruent.
6-106 Construct a triangle congruent to a given triangle.
6-107 Identify: Polygon, quadrilateral, triangle, square, rectangle, parallelogram.
6-108 Find the perimeters of the above polygons.
6-109 Compute the areas of rectangles, squares, parallelograms and triangles, using formulas.
6-110 Find the circumference of a circle.
6-111 Identify: Solid, face, base, cube, cone, cylinder, pyramid, sphere, half-sphere, cubic units, volume.

LEARNING ACTIVITY PACKETS FOR LEVEL 6

OBJECTIVES 6-(71, 72)

MULTIPLY AND DIVIDE BY POWERS OF TEN

OBJECTIVE:

The student will be able to:

Multiply by powers of 10 by moving the decimal point.

Divide by powers of 10 by moving the decimal point.

PRETEST

1. How many zeros are there in each of these powers of ten?
   10^0 10^3 10^5 10^{50} 1,000 1,000,000
2. Write these numbers as powers of 10:
   10 1000 1 1,000,000 10,000
3. Multiply by moving the point:
   By 10: 36 500 .08 16.7 1.096
4. Divide by moving the point:
   By 100: 1500 650 25.96 1.08
5. Make up a rule for multiplying any number by any power of 10.
6. Make up a rule for dividing any number by any power of 10.

Now that you have learned how to work with exponents and decimal fractions, you can multiply and divide easily without doing the work out.

Let's start with multiplication. Of course, you can only use powers of 10 when you use this method.

Also, remember that the decimal point is always to the right of a whole number, but we do not write it unless it is followed by a decimal fraction.

EXAMPLE:

25	=	25.
11,085	=	11,085.
2,058,712	=	2,058,712.

Step 1: Count the number of zeros in the power of 10 by which you are multiplying.
Step 2: Move the decimal point to the right the same number of places as there are zeros in the power of 10.

Example: Multiply 13 by 10^2. The exponent tells us there are 2 zeros.

So, $13. \times 10^2 = 13.00 = 1300.$

EXERCISE:
See, if you can do these:

1) 36 x 1,000
2) 14 x 10^0
3) 3.4 x 10^4
4) 3.478 x 10,000

(Note: Any number to the zero power = 1
$10^0=1$
$36^0=1$
$14{,}235^0=1$)

Are these your answers?

1) 36,000 2) 14 3) 34,000 4) 34, 780

Very Good!

It is easy to divide, too! Division is the opposite of multiplication, so just move the point in the opposite direction. You move the point to the right to multiply. To divide move the point to the left the same number of places as there are zeros in the power of 10.

EXAMPLE:

$35 \div 10^2$ The 2 tells us there are 2 zeros. So,

$35 \div 10^2 =$
$35. \div 100 = .35$

Now, try these:

1) $3000 \div 10^1$
2) $365 \div 10^4$
3) $4568 \div 1000$
4) $362.5 \div 10^5$

Did you get these answers?

1) 300 2) .0365 3) 4.568 4) .003625

POSTTEST

1. Division is the ________________ of multiplication.
2. The exponent in a power of 10 tells you how many ________________ to use.
3. Multiply without doing the work:

a) $10^1 \times 10^1$
b) $10^2 \times 41$
c) 2.89×10^3
d) $.08 \times 10^5$
e) 1000×10^3
f) 89.605×10^1
g) $4.89 \times 10^3 \times 10^0$
h) 16.16×10^2
i) 1085×10^3
j) $.0246 \times 10^4$

4. Divide without doing the work:

a) $10^2 \div 10^2$
b) $1000 \div 100$
c) $468 \div 10^2$
d) $1.234 \div 10^1$
e) $56.23 \div 10^3$
f) $1{,}000{,}000 \div 1000$
g) $100 \div 1000$
h) $10 \div 10^4$
i) $100{,}000 \div 10^5$
j) $.087 \div 100$

OBJECTIVE 6-74
SQUARING A NUMBER

OBJECTIVE:

The student will be able to square any number.

PRETEST

1. How do you read the following?
   2^2 5^2 8^2 1.5^2

2. Square these numbers:
   18 20 32 14.5

An exponent tells you how many times to use a number as a factor.

$$2^2 = 2 \times 2$$

2 is used as a factor two times. (Remember: a factor is one of the numbers you multiply.)

This is how we read 2^2: 2 to the second power or 2 squared.

You can see from this that when we square a number, we multiply the number by itself.

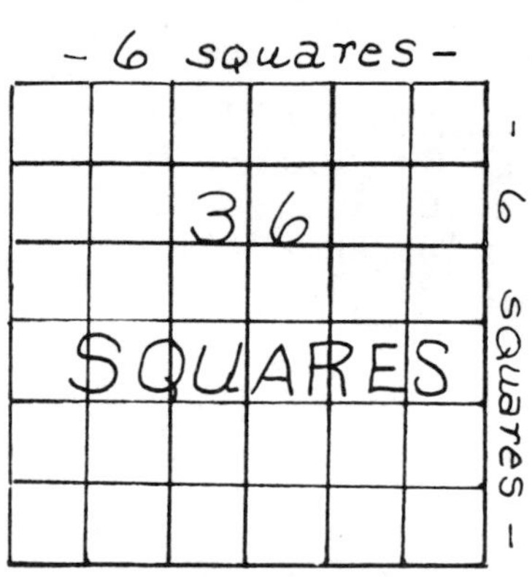

Notice that there are 6 squares on each side of this figure. If you count the squares, you will notice that there are 36 squares inside.

$$6^2 = 6x6 = 36$$

In the same way—
$4^2 = 4x4 = 16$
$9^2 = 9x9 = 81$
$21^2 = 21 \times 21 = 441$

<u>EXERCISE:</u>

Now, <u>you</u> find these squares:

5^2 118^2 500^2 33^2 99^2 24^2

You know how to multiply decimal fractions. These are squared in the same way:

$25.1^2 = 25.1 \times 25.1$

```
multiplicand ———————>   25.1
multiplier ——————————> x 25.1
                        ------
                         251
                        1255
                        502
                        ------
Product —————————————>  630.01
```

Note: Be sure to count the number of places to the right of the decimal point in the multiplicand and multiplier to see where to put the point in the product.

<u>EXERCISE:</u>

Now <u>you</u> find these squares:

13.6^2 21.8^2 100.3^2 1.82^2 $.08^2$

DO YOU WANT TO KNOW <u>an easier way to square a number between 10 and 100 when the ones digit is 5?</u>

Here is how you do it.

<u>EXAMPLE:</u> square 45

Step 1: Multiply the 10s digit
by a number one greater.
5 (1 greater than 4)x4 = 20

Step 2: Square the ones digit.
$5^2 = 5 \times 5 = 25$

Step 3: Write the numeral from step 1. Then write the numeral from step 2.
$2025 = 45^2$

<u>EXERCISE:</u>

Use this method to square the following:

15^2 25^2 35^2 55^2 65^2 75^2 85^2 95^2

<u>POSTTEST</u>

1. Explain how to square any number.
2. Square these numbers without doing the multiplication out: 25 15 75

3. Find these squares:

8	7	4	3	9	6
28	27	44	53	72	81
3.9	4.8	50.1	25.23		

4. Draw a diagram to show that $3^2 = 9$.
5. <u>Optional: to prove you know your stuff!!</u>

Square: 1/3 5/8 3/4 1 1/8

OBJECTIVE 6-79

DIVISIBILITY

OBJECTIVE: The student will be able to use divisibility rules (2,3,4,5,6,8,9,10).

PRETEST

Fill in the blanks with the appropriate numbers.

1. The number__________ divides all even numbers.
2. All numbers are divisible by______.
3. A number is divisible by another number when it has______ for a remainder.

4. The number 15 is divisible by ____, ____, ____ and ____.
5. The prime number 11 is divisible by________ and_____.
6. What is the rule for divisibility by 8?

WHAT IS DIVISIBILITY?

To start this unit we first must know what divisibility is. A number is divisible by another number if it can be divided exactly by that other number with a zero remainder.

Here are some examples:

A. 15 is divisible by 5. (5 divides 15 three times with a zero remainder.)

B. 22 is divisible by 2. (2 divides 22 eleven times with a zero remainder.)

C. 22 is not divisible by 5. (5 divides 22 four times with a remainder of 2.)

What numbers will divide 18?

What numbers will not divide 18?

We must be careful to remember that some numbers may be divisible by one number but not by another.

Here are some examples:

A. 26 is divisible by 2 but not by 3.

B. 64 is divisible by 4 but not by 5.

Some numbers may be divisible by more than one number.

EXAMPLES:

A. 12 is divisible by 1, 2, 3, 4, 6, and 12.

B. 21 is divisible by 1, 3, 7, and 21.

EXERCISE:

1. Give some examples of numbers that are
   a. Divisible by 2.
   b. Divisible by 3.
   c. Divisible by both 3 and 5.
2. Are the following numbers divisible by 3?
   a. 18
   b. 61
   c. 22
   d. 732
   e. 901
   f. 385,611

Sometimes, we can see at a glance what numbers will divide other numbers. For example, we know that 3 will divide 21 but what about larger numbers? What numbers will divide 836,592,183?

We will study and apply a series of tests that will enable us to tell what numbers will divide others.

DIVISIBILITY BY 1

This is the easiest test of all, because all numbers are divisible by one. If a number is divisible by 1 and itself and no other number, we say that it is a prime number. For example, 7 is divisible by 1 and 7, so 7 is a prime number.

EXERCISE:

Which of the following are prime numbers? Circle your answers.

a) 5 b) 9 c) 3 d) 17 e) 49 f) 51

DIVISIBILITY BY 2

A number is divisible by two, if the units digit of the number ends in 0, 2, 4, 6, or 8. (All even numbers are divisible by 2.)

EXAMPLES:

a. 13,492 is divisible by 2.
b. 863, 691,050 is divisible by 2.

EXERCISE:
Give 5 examples of numbers with 3 or more digits that are divisible by 2.

a)
b)
c)
d)
e)

DIVISIBILITY BY 3

A number is divisible by 3 if the sum of the digits is divisible by 3.

EXAMPLES:

A. 2541 is divisible by 3, because 2 + 5 + 4 + 1 =12 and 12 is divisible by 3.
B. 35,142 is divisible by 3, because 3 + 5 + 1 + 4 + 2 =15 and 15 is divisible by 3.
C. 1526 is not divisible by 3, because 1 + 5 + 2 + 6 = 14 and 14 is not divisible by 3.

EXERCISE:
Apply the test for divisibility by 3 to determine which of the following are divisible by 3. Circle your answers.

a) 558   b) 7,213   c) 6,087   d) 46, 881

DIVISIBILITY BY 4

A number is divisible by 4 only if it is an even number and the number represented by the last two digits (tens and units) is divisible by 4. Numbers ending in two zeros are divisible by 4.

EXAMPLES:

A. 58,264 is divisible by 4, because 64 is divisible by 4.
B. 27, 828 is divisible by 4, because 28 is divisible by 4.
C. 5300 is divisible by 4.
D. 4,634 is not divisible by 4.

EXERCISE:

Using the test for divisibility by 4, circle those numbers that are divisible by 4:

a) 832 b) 6,794 c) 8,500 d) 71, 676

DIVISIBLITY BY 5

A number is divisible by 5 only if it ends in 5 or 0.

EXAMPLES:

A. 8,325 ends in 5, so it is divisible by 5.
B. 59,740 ends in 0, so it is divisible by 5.

EXERCISE:

Circle those numbers that are divisible by 5.

a) 678 b) 9,000 c) 5,265 d) 85, 305

DIVISIBILITY BY 6

A number is divisible by 6, if it is divisible by both 2 and 3.

EXAMPLES:

A. 156 is divisible by both 2 and 3, so it is divisible by 6. Check this by applying the divisibility tests for 2 and 3.
B. 536 is divisible by 2 but it is not divisible by 3, so it is not divisible by 6.

EXERCISE:

Circle those numbers that are divisible by 6:
a) 393 b) 588 c) 1,968 d) 185,202

DIVISIBILITY BY 8

A number is divisible by 8 only if it is an even number and the number represented by the last three digits is divisible by 8.

EXAMPLES:

A. 1,888 is divisible by 8 because 888 is divisible by 8.
B. 10,648 is divisible by 8 because 648 is divisible by 8.
C. 256,906 is not divisible by 8 because 906 is not divisible by 8.

EXERCISE:

Circle those numbers that are divisible by 8:
a) 1,532 b) 23,824 c) 1,085,568 d) 356,872

DIVISIBILTY BY 9

The divisibility rule for 9 is the same as for 3. A number is divisible by 9 if the sum of the digits is divisible by 9.

EXAMPLES:

A. 36,927 is divisible by 9 because 3 + 6 + 9 + 2 + 7 =27 and 27 is divisible by 9.
B. 126,368 is not divisible by 9, because 1 + 2 +6 + 3 + 6 +8 =26 and 26 is not divisible by 9.

EXERCISE:

Circle those numbers that are divisible by 9:
a) 437,634 b) 639,936 c) 927,999 d) 1,087

DIVISIBILITY BY 10

Any number that ends in zero is divisible by 10.

EXAMPLES:

1,23<u>0</u> 630,89<u>0</u> 40,23<u>0</u> 1,087,00<u>0</u> 98,76<u>0</u> 320,92<u>0</u>

EXERCISE:

Circle those numbers that are divisible by 10:
a) 28,942 b) 34,000 c) 32,678 d) 100,000

POSTTEST

Using the tests for divisibility of numbers 1 through 4, tell which numbers the following are divisible by:
a) 83,592 b) 578,976 c) 67,500 d) 468,763

Determine whether the following numbers are divisible:

by 4:	6,314	332	9,000	8,121	91,836
by 3:	311	93,111	8,380	6,221	501
by 5:	1,234	16,505	12,895	46,222	123,454
by 6:	624	6,186	15,988	42,654	123,963
by 8:	5,888	7,108	96,968	35,862	327,728
by 9:	3,456	9,639	46,352	172,349	999,981
by 10:	2,500	18,320	10,100	180,320	595,125

Give two examples of numbers with 5 or more digits that are:
a) divisible by 2.
b) divisible by 5.
c) divisible by both 5 and 2.
d) divisible by both 3 and 4.

OBJECTIVES 6-(80-83)
BINARY (BASE 2) NUMERATION

<u>OBJECTIVES:</u> The student will be able to:

State the place value of numerals in Base 2 up to the sixth place.
Change numerals from Base 2 to Base 10.
Change numerals from Base 10 to Base 2.
Write numerals up to 50 in Base 2.

PRETEST

1. What is the value of the digit in the third place in Base 10? In Base 2?
2. Change 27_{10} to Base 2
3. Change 11_2 to Base 10
4. How do you read: 10111_2?

HOW CAN THIS BE?
2 + 3 = 101
20 - 2 = 10010
16 ÷ 2 = 1000
7 x 8 = 111000
Hint: What base is used for the answers?

If you decided that the above examples are written in Base 10 (the decimal system) and the answers are written in Base 2, you are right!

This is how the answers should be written so we can tell what base is being used: 101_2 or 101 (Base two), 10010_2 or 10010 (Base two), 1000_2 or 1000 (Base 2), 111000_2 or 111000 (Base 2).

WHEN WE WORK WITH ANY NUMBER BASE OTHER THAN 10, WE MUST IDENTIFY THE BASE.

In Base 10, we use 10 digits (0,1,2,3,4,5,6,7,8,9)

In Base 3, we use 3 digits (0,1,2,)

How many digits are used in base 2?

Notice that we do not use the digit of the base we are using. We do not use 5 in Base 5, 10 in Base 10, etc.

Can you explain why? (If not, ask the teacher)

In Base Two, each place is worth twice as much as the place to the right. Each place is worth one-half as much as the place to the left.

A PLACE VALUE table for BASE TWO would look like this: (6 places)

Thirty-twos	Sixteens	Eights	Fours	Twos	Ones
32	16	8	4	2	1

HOW TO CHANGE A NUMERAL FROM BASE 2 TO BASE 10

EXAMPLE: Change 11010_2 to Base Ten

Step 1: Expand the numeral:

$11010_2 = (1 \times 16) + (1 \times 8) + (0 \times 4) + (1 \times 2) + (0 \times 1)$

Add: $16 + 8 + 0 + 2 + 0 = 26$

So, $11010_2 = 26_{10}$

This is read:

One one zero one zero Base Two. Notice that we do not read numerals in other bases the same way that we read them in Base 10. We do not have a "sixteen" in any base except 10. In another base it would be read "one six."

EXAMPLE: Change 1111_2 to Base 10.

$1111_2 = (1 \times 8) + (1 \times 4) + (1 \times 2) + (1 \times 1)$

$1111_2 = 8_{10} + 4_{10} + 2_{10} + 1_{10} = 15$

These little numbers are called: SUBSCRIPTS.

Now you try it!

Change these Base Two numerals to Base Ten:

10_2 1011_2 11_2 111_2 1001_2 101_2 1101_2

ARE YOU ASKING: "WHY SHOULD I LEARN THIS?"

One reason is to show you there are number systems other than the one we are used to working with. Working with another system will help you to understand our system better.

Also, you are interested in computers, aren't you? Binary (Base Two) numerals are used in digital computers. A two-digit system makes it possible to either open or close electrical circuits: 1=close, 0=open.

HOW TO CHANGE A NUMERAL FROM BASE 10 TO BASE 2:

EXAMPLE: Change 29_{10} to Base 2.

Step 1: Look at the place value table. Find the largest value that can be divided into 29. The largest one is 16.

Step 2: Determine how many digits the Base 2 numeral will have by seeing what place 16 holds. It is in the 5th place. Write 5 dashes (_ _ _ _ _) so you will remember.

Step 3: Divide 29 by 16⟶1 with a remainder of 13. Put a 1 over the first dash (1 _ _ _ _). You have to find 4 more digits.

Step 4: Take the remainder, 13, and repeat step 3, using the next smaller place value. $13 \div 8 = 1$ with a remainder of 5. (1 1 _ _ _)

$5 \div 4 = 1$ with a remainder of 1

(1 1 1 _ _)

1 is not divisible by 2, so (1 1 1 0 _)

Put the remainder in the ones place (1 1 1 0 1)

Answer: $29_{10} = 11101_2$

EXERCISE:

Now, you are ready to use the place value table to get the numbers from 1-50.

$1 = 1_2$ $\quad$ $3 = 11_2$

$2 = 10_2$ $\quad$ $4 = 100_2$

Now finish the list.
Notice that 1 is always in any system.

POSTTEST

1. How many digits in any base?
2. Another name for Base 2 is the B_________ system.
3. Write in Base 2:
   17 63 47 55 38 22
4. Write in Base 10:
   a) 111_2 b) 1011_2 c) 10001_2 d) 111111_2 e) 11111_2

CHALLENGE: What would be the place values of the 4 places to the left of the thirty-twos place?

OBJECTIVE 6-84

BINARY ADDITION AND SUBTRACTION

<u>OBJECTIVE:</u> The student will be able to add and subtract in Base 2.

<u>PRETEST</u>

1. Change these Base 10 numerals to Base 2:
   a) 55 b) 39 c) 43
2. Change these Base 2 numerals to Base 10:
   a) 111_2 b) 10011_2 c) 100110_2
3. Make a table showing the first 6 place values of Base 2.
4. Add: $\begin{array}{r} 11_2 \\ +111_2 \\ \hline \end{array}$ $\quad$ $\begin{array}{r} 101110_2 \\ +111110_2 \\ \hline \end{array}$

5. Subtract:

   $\begin{array}{r} 111_2 \\ -101_2 \\ \hline \end{array}$ $\quad$ $\begin{array}{r} 10001_2 \\ -1001_2 \\ \hline \end{array}$

Adding in Base 2 <u>will not be hard</u> if you remember to carry groups of 2 just as you carry groups of 10 when working in Base 10. You cannot write a 2 in Base 2, just as you cannot write a 10 in Base 10. The only digits you can use are 0, 1.

<u>EXAMPLE:</u> Adding in Base 10: 1 + 1 =2

Adding in Base 2: $1_2 + 1_2 = 10_2$

<u>This is the hardest binary addition fact you will have to learn.</u>

<u>3 OTHER BINARY ADDITION FACTS</u>

$0_2 + 0_2 = 0_2$
$0_2 + 1_2 = 1_2$
$1_2 + 0_2 = 1_2$

PERFORM THIS ADDITION:

$$\begin{array}{r} 10101_2 \\ 1000_2 \\ \hline 11101_2 \end{array}$$

This was easy because none of the sums were 2. They all added up to 1.

Whenever the answer equals 2, we write 0 in the sum and carry 1 to the next place.

Study these examples:

$$\begin{array}{r} 11_2 \\ +10_2 \\ \hline 101_2 \end{array} \qquad \begin{array}{r} 111_2 \\ +101_2 \\ \hline 1100_2 \end{array}$$

Check these answers by changing the numerals to Base 10 and adding.

EXERCISE:

a) $\begin{array}{r} 10_2 \\ 11_2 \\ 11_2 \\ \hline \end{array}$ b) $\begin{array}{r} 111_2 \\ +10_2 \\ \hline \end{array}$ c) $\begin{array}{r} 1100_2 \\ +1011_2 \\ \hline \end{array}$ d) $\begin{array}{r} 1110_2 \\ +1011_2 \\ \hline \end{array}$ e) $\begin{array}{r} 10110_2 \\ 1111_2 \\ 10101_2 \\ \hline \end{array}$

SUBTRACTION IN BASE 2

4 BINARY SUBTRACTION FACTS

$0_2 - 0_2 = 0_2$
$1_2 - 0_2 = 1_2$
$1_2 - 1_2 = 0_2$
$10_2 - 1_2 = 1_2$

To "borrow", you will use the 4th fact.

When you do not have to "borrow" it is easy to do binary subtraction.

$$\begin{array}{r} 11_2 \\ -\ 0_2 \\ \hline 11_2 \end{array} \qquad \begin{array}{r} 111_2 \\ -\ 11_2 \\ \hline 100_2 \end{array} \qquad \begin{array}{r} 1011_2 \\ -1001_2 \\ \hline 10_2 \end{array} \qquad \begin{array}{r} 110_2 \\ -110_2 \\ \hline 0_2 \end{array}$$

LOOK AT THIS PROBLEM

$$\begin{array}{rl} 110_2 & \text{(minuend)} \\ -\ 1_2 & \text{(subtrahend)} \\ \hline \end{array}$$

In this problem, the ones place has a bigger digit in the subtrahend than in the minuend.

$$\begin{array}{r} {\scriptstyle 0\,1\,0} \\ 1\not{1}0_2 \\ -\ 1_2 \\ \hline \end{array}$$

We must regroup before we can subtract. The 1 in the twos place becomes 0. The 0 in the ones place becomes 10_2. (Remember: 10_2=2). You are really changing one 2 into 2 ones.

$$\begin{array}{r} {\scriptstyle 0\,1\,0} \\ 1\not{1}0_2 \\ -\ 1_2 \\ \hline 101_2 \end{array}$$

Check:

$$\begin{array}{llr} 110_2 & = 4 + 2 + 0 = & 6_{10} \\ -\ 1_2 & = & -\ 1_{10} \\ \hline 101_2 & = 4 + 0 + 1 = & 5_{10} \end{array}$$

It checks.

Study this example:

$$\begin{array}{lll} 1101_2 & = 8+4+0+1 & = 13_{10} \\ -\ 10_2 & -\ 2+0 & -\ 2_{10} \\ \hline 1011_2 & 8+0+2+1 & 11_{10} \end{array}$$

PRACTICE—CHECK BY CHANGING TO BASE 10 AND SUBTRACTING OR BY BINARY ADDITION

EXAMPLE:

$$\begin{array}{lll} 11010_2 = 16+8+0+2+0 = 26_{10} & \text{OR} & 10011_2 \\ -\ 111_2 = \quad 4+2+1 \qquad\quad -\ 7_{10} & & +\ 111_2 \\ \hline 10011_2 \qquad\qquad\qquad\qquad\quad 19_{10} & & 11010_2 \end{array}$$

EXERCISE

a) $\begin{array}{r} 1110_2 \\ -\ 101_2 \\ \hline \end{array}$ b) $\begin{array}{r} 11010_2 \\ -\ 1100_2 \\ \hline \end{array}$ c) $\begin{array}{r} 11010_2 \\ -\ 1000_2 \\ \hline \end{array}$ d) $\begin{array}{r} 1101_2 \\ -\ 110_2 \\ \hline \end{array}$ e) $\begin{array}{r} 1000_2 \\ -\ 100_2 \\ \hline \end{array}$

POSTTEST

1. Add and check by changing to Base 10:

a) $\begin{array}{r} 1111_2 \\ +1111_2 \\ \hline \end{array}$ b) $\begin{array}{r} 1010_2 \\ +\ 101_2 \\ \hline \end{array}$ c) $\begin{array}{r} 10_2 \\ 110_2 \\ 1110_2 \\ \hline \end{array}$ d) $\begin{array}{r} 110_2 \\ 1110_2 \\ 11_2 \\ \hline \end{array}$

Challenge: e) $\begin{array}{r} 110110_2 \\ +101111_2 \\ \hline \end{array}$

2. Subtract and use either method to check;

(Change to Base 10 or check by addition .)

a) $\begin{array}{r} 111_2 \\ -\ 11_2 \\ \hline \end{array}$ b) $\begin{array}{r} 100011_2 \\ -\ 10001_2 \\ \hline \end{array}$ c) $\begin{array}{r} 11001_2 \\ -\ 10010_2 \\ \hline \end{array}$ d) $\begin{array}{r} 1110_2 \\ -\ 101_2 \\ \hline \end{array}$ e) $\begin{array}{r} 100_2 \\ -\ 11_2 \\ \hline \end{array}$

OBJECTIVE: 6-85

BINARY MULTIPLICATION AND DIVISION

OBJECTIVE: The student will be able to multiply and divide in Base 2.

PRETEST

1. Multiply:

   a) $11_2 \times 10_2$ b) $1101_2 \times 11_2$ c) $11101_2 \times 111_2$

   d) $1110_2 \times 101_2$ e) $1001_2 \times 1001_2$

2. Divide:

   a) $110_2\,\overline{)111010_2}$ b) $11_2\,\overline{)1011_2}$

BINARY MULTIPLICATION

Binary multiplication uses the same method as decimal multiplication.

We need to remember 4 basic facts. (The subscript is all that is different.)

$0_2 \times 0_2 = 0_2$
$0_2 \times 1_2 = 0_2$
$1_2 \times 0_2 = 0_2$
$1_2 \times 1_2 = 1_2$

Study this example:

$$
\begin{array}{rl}
1101_2 & \text{multiplicand} \\
\times\ 110_2 & \text{multiplier} \\
\hline
0000 & \\
1101 & \\
1101 & \\
\hline
1001110_2 &
\end{array}
$$

0000, 1101, 1101 → partial products (Write the zeros, if you find it easier.)

The hard part is remembering to "carry" when the partial products add up to 2.

Indent partial products in binary multiplication just as you do in decimal multiplication.

FILL IN these partial products:

a) 1110_2	b) 111_2	c) 1101_2
x 101_2	x 101_2	x 110_2
----	---	----
----	---	----
----	---	----
1000110_2	100011_2	1001110_2

Note: Keep partial product columns nice and straight. It will be easier to do.

When there are more than two partial products, it is easier to use the partial sums method.

$$
\begin{array}{rl}
11010_2 & \\
\times\ 111_2 & \\
\hline
11010 & \leftarrow \text{first partial product} \\
11010 & \leftarrow \text{second partial product} \\
\hline
1001110 & \leftarrow \text{partial sum} \\
11010 & \leftarrow \text{third partial product} \\
\hline
10110110_2 & \leftarrow \text{final product}
\end{array}
$$

Check: $\begin{array}{r} 26 \\ \times\ 7 \\ \hline 182_{10} \end{array}$

Does $10110110_2 = 182_{10}$?

EXERCISE: Use the partial sums method:

a) $\begin{array}{r} 1000_2 \\ \times\ 111_2 \\ \hline \end{array}$ b) $\begin{array}{r} 1110_2 \\ \times\ 1100_2 \\ \hline \end{array}$ c) $\begin{array}{r} 10000_2 \\ \times\ 101_2 \\ \hline \end{array}$ d) $\begin{array}{r} 111_2 \\ \times\ 111_2 \\ \hline \end{array}$

Check: Change Base 2 numerals to Base 10 and multiply.

BINARY DIVISION

There are 2 Binary Division facts:

$1_2 \div 1_2 = 1_2$
$0_2 \div 1_2 = 0_2$

Study this example:

$$\begin{array}{r|l} & \quad 1110_2 \\ \hline 100_2 & 111010_2 \\ & -100 \\ \cline{2-2} & \quad 110 \\ & \quad 100 \\ \cline{2-2} & \quad\ 101 \\ & \quad\ 100 \\ \cline{2-2} & \quad\quad 10 \end{array}$$

Remainder $= 10_2$

Notice that binary division follows the method for decimal division.

EXERCISE:

a) $110_2 \overline{)100010_2}$ b) $110_2 \overline{)111010_2}$ c) $11_2 \overline{)1011_2}$

Check by binary multiplication.

POSTTEST

1. Multiply:

a) $\begin{array}{r} 101_2 \\ \times\ 111_2 \\ \hline \end{array}$ b) $\begin{array}{r} 100_2 \\ \times\ 110_2 \\ \hline \end{array}$ c) $\begin{array}{r} 111_2 \\ \times\ 111_2 \\ \hline \end{array}$ d) $\begin{array}{r} 1110_2 \\ \times\ 101_2 \\ \hline \end{array}$ e) $\begin{array}{r} 110110_2 \\ \times\ 1101_2 \\ \hline \end{array}$

2. Divide:

a) $11_2 \overline{)11111_2}$ b) $10_2 \overline{)10110_2}$ c) $110_2 \overline{)110000_2}$

d) $101_2 \overline{)1001011_2}$ e) $101_2 \overline{)110111_2}$

OBJECTIVES 6-(86-88)
LINEAR METRIC MEASURE

OBJECTIVES: The student will be able to:
Define the prefixes used in the Metric System.
Express metric measures as powers of 10.
Use linear measure in the Metric System.

PRETEST

1. What do the following prefixes mean?
a) deci- b) centi- c) milli- d) deca- e) hecto- f) kilo-

2. Which is longer, a meter or a yard? How much longer?
3. Express the prefixes used above as powers of 10.

There are other systems of measure besides the one you use. You are going to learn about the Metric System. The Metric System of measure is used by scientists in our country and by everybody in many other countries. The Metric System was invented in the 18th century by some French scientists who believed that a system of measurement based on the decimal system of numeration would be easier and more convenient to use.

WHY YOU SHOULD LEARN ABOUT IT. The United States has a Commission working on ways to get the Metric System in use here. You will see it already in use in many places if you start looking around. Patterns for making clothes have both the English system (the one we use most now) and the Metric System of measurement; machinery parts; parts for cars; and items in the grocery stores. You will want to be ready when, and if, the Metric System comes into full-scale use in the United States.

STUDY AND LEARN THE FOLLOWING PREFIXES:

Milli- means one thousandth, so a millimeter is one thousandth of a meter.

Centi- means one hundredth, so a centimeter is one hundredth of a meter.

Deci- means one tenth, so a decimeter is one tenth of a meter.

THE METER IS THE BASIC UNIT OF MEASURE IN THE METRIC SYSTEM. A METER IS 3.37 INCHES LONGER THAN A YARD, SO A METER IS 39.37 INCHES LONG. Get a yardstick and a meter stick and compare them.

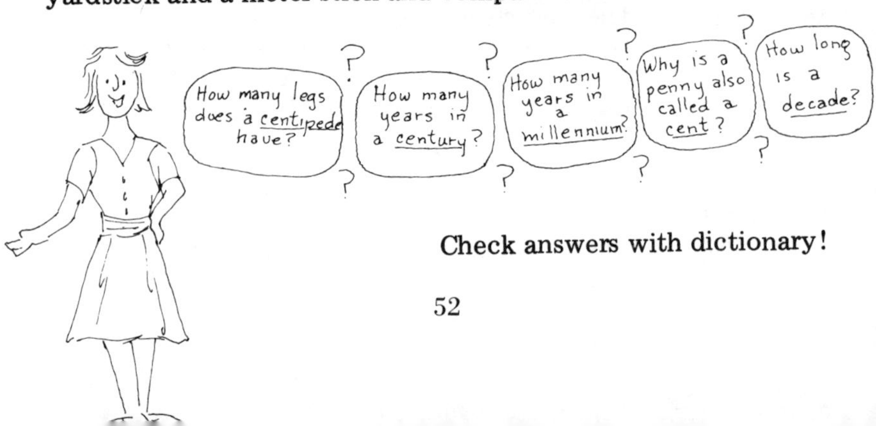

Check answers with dictionary!

HAVE YOU LEARNED THE MEANING (AND SPELLING) OF THE METRIC PREFIXES? IF YOU HAVE NOT LEARNED THEM, DO NOT GO ON UNTIL YOU DO!

Get a meter stick. Examine it closely. Find the beginning and the end of the biggest sections into which the meter stick is divided. You will see that there are 10 of these large divisions. Each one of these divisions is called a decimeter, because it is equal to one tenth of a meter.

Now, find the ten parts into which the decimeter is divided. Each one of these divisions is called a centimeter. A centimeter is one tenth of a decimeter. Do you also see that a centimeter is one hundredth of a meter?

A centimeter is divided into 10 very small parts. Each of these divisions is called a millimeter. A millimeter is one tenth of a centimeter. Do you also see that a millimeter is one thousandth of a meter?

STUDY THESE RELATIONSHIPS UNTIL YOU UNDERSTAND THEM, THEN SEE IF YOU CAN DO THE FOLLOWING EXERCISES.

1. How many centimeters are there in a decimeter?
2. How many centimeters are there in a meter?
3. How many decimeters are there in a meter?
4. How many millimeters are there in a decimeter?
5. How many millimeters are there in a centimeter?
6. How many millimeters are there in a meter?

Get a meter stick and find the perimeter (the distance around) your classroom. (State the measurement in meters.) If the measurement does not come out even, how many decimeters are there?

Measure a desk top and find how many decimeters, centimeters and millimeters it measures.

Choose another item and measure it in the appropriate units.

Report the measurements you have found on the above three items to your teacher.

NOW, YOU SHOULD BE READY TO LEARN THE NEXT THREE PREFIXES:

Deca- means ten, so a decameter is ten meters.

Hecto- means one hundred, so a hectometer is one hundred meters.

Kilo- means one thousand, so a kilometer is one thousand meters.

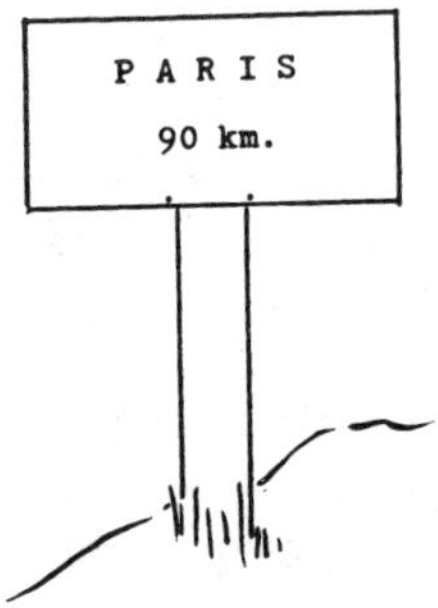

EXERCISE: If you were in Europe and saw this sign, would you have to travel more or less than 90 miles to get to Paris? Km. is the abbreviation for kilometer.

To answer this question, you will have to find what part a kilometer is of a mile.

Look up kilometer in the Dictionary.

Now, let's see how well you understand the second set of prefixes you have learned.

EXERCISE:

1. Name three units of metric measure that are longer than the meter.
2. Which is longer, a kilometer or a hectometer?
3. Which is longer, a millimeter or a kilometer?
4. In the Olympic Games there is an event called the 50,000-meter walk. How many kilometers is this?

CHALLENGE: It is approximately 30 miles from Fitchburg to Worcester. How many kilometers is that?

ONE OF THE REASONS THAT THE METRIC SYSTEM OF MEASUREMENT IS CONVENIENT TO USE IS THAT THE RELATIONSHIPS BETWEEN THE VARIOUS UNITS OF MEASURE CAN BE EXPRESSED AS POWERS OF 10.

You will remember from previous work you have done with exponents that you can tell the number of zeros to write when using 10 as a base by looking at the exponent. 10^2 means two zeros or 100, 10^3 means three zeros or 1000, etc. Likewise 10^0 means no zeros or just 1. (Also, any number raised to the zero power = 1.)

Since the meter is the basic unit of measure in the Metric

System, we let 10^0 = the meter. Every other unit is either bigger or smaller than the meter.

The next larger unit is the decameter, so 10^1 represents the decameter. $10^1 = 10$. (Any number raised to the first power = the number itself.)

The next larger unit to the decameter is the hectometer, so 10^2 represents the hectometer. $10^2 = 100$.

The next larger unit to the hectometer is the kilometer, so 10^3 represents the kilometer. $10^3 = 1000$.

Let's put it all together:

$10^0 = 1$ = the basic unit = the meter.
$10^1 = 10$ = the decameter.
$10^2 = 100$ = the hectometer.
$10^3 = 1000$ = the kilometer.

Likewise:

10 meters = 1 decameter
100 meters = 1 hectometer
1000 meters = 1 kilometer

MAKE SURE YOU UNDERSTAND ALL OF THE FACTS GIVEN ABOVE BEFORE GOING ON!

All of the relationships shown above deal with units that are larger than the meter. Now, we are going to study how to express relationships that are smaller than the meter by using powers of 10.

The next smaller unit to the meter is the decimeter. You saw that units that are smaller than the meter are expressed as fractions. The decimeter is 1/10 of the meter. The exponent that we used with the meter was 0. What is the next number smaller than 0? If you said -1, you were correct. You can check this by looking at the number line in your classroom. So, you can see that 10^{-1} represents the decimeter.

The next smaller unit to the decimeter is the centimeter, so we use the next smaller exponent. What is it? Look at the number line. As you go to the left on the number line the numbers get smaller. Do you see that the next smaller number is 10^{-2}? 10^{-2} represents the centimeter. The centimeter is 1/10 of the decimeter.

The next smaller unit to the centimeter is the millimeter; 10^{-3} represents the millimeter. The millimeter is 1/10 of the centimeter.

From this explanation, you can see that:

10^{-1} = 1/10 = the decimeter.
10^{-2} = 1/10 x 1/10 = 1/100 = the centimeter.
10^{-3} = 1/10 x 1/100 = 1/1000 = the millimeter.

REVIEW THE MEANINGS OF THE PREFIXES AND HOW TO EXPRESS METRIC MEASURES USING POWERS OF 10. WHEN YOU FEEL YOU ARE READY, GO ON TO THE POSTTEST.

POSTTEST

1. Give the meanings of the following prefixes used in the Metric System:
   a) hecto- b) deci- c) deca- d) kilo- e) milli- f) centi-
2. Write the powers that are used with each of the above.
3. Which is larger?
   a) Centimeter or decimeter
   b) Hectometer or decameter
   c) Millimeter or kilometer
4. Give fractions to represent the decimeter, centimeter and millimeter.
5. Would you like to work with the Metric System all of the time instead of the English system which you now use? Why or why not?

OBJECTIVES 6- (91, 92)

FINDING A PERCENT OF A NUMBER

OBJECTIVES: The student will be able to:
Express ratios as parts of 100 and change to percents.
Use a proportion to find a percent of a number.

PRETEST

1. Which of the following represent ratios?
a) 3.14 b) 1/2 c) 1/4 d) 3:2 e) 4 to 5
2. Change these decimals to fractions with powers of 10 in the denominators:
a) .05 b) .8 c) .009 d) .07
3. Change these fractions to percents:
a) 1/2 b) 1/4 c) 20/100
4. What is a proportion?

We use a ratio to compare two numbers by division. We can express ratios in several different ways.

EXAMPLE: There are 25 pupils in the mathematics class. Eleven of the students are girls. What is the ratio of girls to boys in the class?

If there are 11 girls, there must be 14 boys. We can show the ratio of girls to boys in these ways:

11 to 14

11 : 14

$$\frac{11}{14}$$

What would the ratio of boys to girls be? Of girls to the number of pupils in the class? Of boys to the number of pupils in the class?

In this packet, you will learn to work with percents. Percent is the ratio of a given number to 100.

Example: 25/100 is what percent?
Answer: 25%

You can see that once a number is over 100, it is very easy to state the percent. All you have to do is use the number in the numerator and add a percent sign.

EXERCISE: Change to percents:

a) 18/100 b) 75/100 c) 12 1/2/100 d) .03 e) .07

R	R	R	A	A	T	I	I	O	T
A	R	R	A	T	I	O			R
T		R	A	T	I	O			A
I		R	R	A	T	I	O		T
O	R	R		R	A	T	I	O	I
	A			A	R	A	T	I	O
	T	R		T		R	I	O	T
	I	R		I			R	R	R
	O	R		O				R	R
	R	A	T	I	O		R	A	

Look at the grid in the illustration. It has 100 small squares. Count the number of times each letter appears in a square and answer the following questions:

1. How many Rs are there? Make a fraction with this number to show the ratio of the number of Rs to the number of small squares in the grid. Change this fraction to a percent.
2. Count the As and do the same thing.
3. Count the Ts and do the same thing.
4. Count the Is and do the same thing.
5. Count the Os and do the same thing.
6. What is the ratio of the empty spaces to the number of small squares in the grid? Change this ratio to percent.

If you did this correctly, you should be able to add up all of the percents you obtained in 1,2,3,4,5, and 7 and get 100%. The ratio named by 100% is equal to 100/100 and is equal to 1.

THE WHOLE OF ANYTHING IS 100% OF IT.

When a fraction does not have a denominator of 100, you must change to an equivalent fraction that does have a denominator of 100. This is easy, if the denominator of the fraction divides evenly into 100.

EXAMPLES:

1/2 does not have a denominator of 100. Two goes evenly into 100. One hundred divided by 2=50. 50/50 is another form of 1. Multiplying both terms of the fraction by 50 does not change its value. It gives you an equivalent fraction. 1 x 50 = 50; 2 x 50 = 100, so 1/2 = 50/100.

EXERCISE:

Change these fractions to equivalent fractions with denominators of 100. Then, change the fractions to percents.

1) 4/5   2) 3/20   3) 4/25   4) 8/10   5) 7/50

If a fraction does not divide evenly into 100, we can put it in fraction form anyhow. We simply express the remainder as a fraction and multiply both parts of the fraction by that mixed number.

EXAMPLE: 3/8

Divide 8 into 100 = 12 1/2. $\frac{12\ 1/2}{12\ 1/2} = 1.$

Multiply 3 x 12 1/2 and multiply 8 x 12 1/2.
Your fraction becomes $\frac{37\ 1/2}{100}$ which = 37 1/2%.

AFTER YOU HAVE STUDIED THESE EXAMPLES CAREFULLY, DO THE FOLLOWING THE SAME WAY:

EXERCISE:

a) 7/8 b) 1/3 c) 5/6 d) 2/9 e) 3/11

PROPORTION

A proportion is a statement that two ratios are equal.
2/3 = 8/12 is a proportion.

We can tell whether two ratios are in proportion to each other either by determining whether they are equivalent ratios or by cross-multiplication.

You have already learned how to find equivalent ratios, so we will go on to the cross-multiplication method.

Multiply the numerator of the first ratio by the denominator of the second ratio. 2 x 12 = 24. Then, multiply the numerator of the second ratio by the denominator of the first ratio. 8 x 3 = 24. 24 = 24, so it is a proportion.

ANOTHER EXAMPLE:

$\frac{1}{2} \times \frac{24}{100}$ Multiply 1 x 100 and 24 x 2. This gives us 100 = 48. This is not a true statement, since 100 =100.

We can not use an equal sign, so we must either use an equal sign with a slash through it or a greater than sign: $\neq$ or $>$.

$1/2 \neq 24/100$ (1/2 is not equal to 24/100)
$1/2 > 24/100$ (1/2 is greater than 24/100)

THEREFORE: This statement is not a proportion; it is an inequality.

USING PROPORTION TO FIND A PERCENT OF A NUMBER

If 20% of the students in a class of 30 wear glasses, how many wear glasses?

To solve this problem, we must first change the percent to a fraction:

$$20\% = 20/100$$

Then, you must find the number that represents 100% of the thing you are trying to find. 100% of the students in the class =30. We will let N (or some other symbol) stand for what we are trying to find. We are trying to find the number of students who wear glasses.

NOW, MAKE A RATIO OF THE NUMBER OF STUDENTS WHO WEAR GLASSES TO THE NUMBER OF STUDENTS IN THE CLASS: N/30

WE ARE NOW READY TO SET UP A PROPORTION:

$$\frac{20}{100} = \frac{N}{30}$$

SOLVE THE EQUATION:

DO THIS BY CROSS-MULTIPLICATION. MULTIPLY 20 x 30

20 x 30 = 600
MULTIPLY N x 100 = 100N
so, 600 = 100N (Divide both sides by 100 to get N.)
6 = N

Answer: 6 students in the class wear glasses.

Study the above example until you understand it. If there is something that you do not understand, ask your teacher to explain it to you. When you are ready, try the following problems:

EXERCISE:

1. If there are 1000 pupils in a school and 24% ride to school on the bus, how many students come by bus?
2. If it rained 6% of the days in April (30 days), how many days were rainy?
3. If a baseball team played 10 games and won 2/5 of the games, how many games did the team win? (Change 2/5 to a percent.)
4. If you bought 1 dozen oranges and 1/3 of them were spoiled when you opened the package, how many oranges were not spoiled?
   (Change 1/3 to a percent. Do not solve this problem by multiplying by 1/3. The reason you are doing these problems is to get practice changing fractions to percents and setting up and solving proportions.)

POSTTEST

1. Change the following percents to fractions with denominators of 100:

25% 1 1/2% 19% 83 1/3% 100% 250% 37 1/2% 3% 103%

2. Change the following percents to fractions and reduce them to lowest terms:

50% 40% 15% 75% 80% 20% 35% 85%

3. What is a ratio?
4. What is a proportion?
5. Explain how to cross-multiply to solve a proportion.
6. Set up proportions and do the following problems:

   a) One hundred dinners were served at a banquet. There was a choice of roast beef or chicken. Thirty-eight percent of the guests ordered chicken. How many ordered roast beef? (Hint: All of a thing always equals 100%.)

b) There was a 20% off sale on clothing. A girl bought a blouse that was marked $9. How much was taken off the original price?

c) A catcher's glove that cost $35 was marked down 10%. How much was the markdown?

OBJECTIVE 6-93

FINDING WHAT PERCENT ONE NUMBER IS OF ANOTHER

OBJECTIVE:

The student will be able to use a proportion to find what percent one number is of another.

PRETEST

Use a proportion to find each of the following percents:

a) 5 is what percent of 10?
b) 100 is what percent of 25?
c) 21 is what percent of 300?
d) 80 is what percent of 500?

Before starting this packet, be sure you have had some practice in solving proportions.

Now you are going to learn how to find the percent when it is missing in a problem. The percent is the number which has the percent sign (%).

EXAMPLE: Find what percent 50 is of 500: First find the number which is equal to 100%. This number is called the base. In this problem, the base is 500.

Make a ratio with the two numbers that are given: $\frac{50}{500}$

We do not know what the percent is. That is what we are looking for in the problem. So, let N (or some other symbol) take the place of the percent until we find out what it is. The ratio for the percent will be represented by $\frac{N}{100}$

Make a proportion with the two ratios:

$$\frac{N}{100} = \frac{50}{500}$$

N x 500 = 500N
100 x 50 = 5000
500N = 5000 (Divide both sides by 500)
N = 10

Replace N with 10: $\frac{10}{100} = 10\%$

Answer: 50 is 10% of 500

ANOTHER EXAMPLE:

15 is what percent of 75?
75 is the base (100% of the number)

Make a ratio: $\frac{15}{75}$

Make a ratio for the percent: $\frac{N}{100}$

Set up a proportion: $\frac{N}{100} = \frac{15}{75}$

Multiply N x 75 = 75N
Multiply 100 x 15 = 1500
75N = 1500 (Divide both sides by 75)
N = 20

$$\frac{N}{100} = \frac{20}{100} = 20\%$$

Answer: 15 is 20% of 75.

STUDY THE PRECEDING TWO EXAMPLES UNTIL YOU UNDERSTAND HOW TO FIND WHAT PERCENT ONE NUMBER IS OF ANOTHER BY SOLVING A PROPORTION. IF YOU WERE UNABLE TO DO THE PRETEST, GO BACK NOW AND TRY TO DO IT. THEN, GO ON TO THE PROBLEMS.

EXERCISE:

1. Kathy's mother baked 5 dozen cookies. Three dozen of the cookies were peanut butter cookies and two dozen were molasses cookies. What percent of the cookies were peanut butter? What percent were molasses?

2. John bought 10 packages of sugar-free gum. He gave 8 packages to his friends and kept 2 packages for himself. What percent of the packages did he give to his friends? What percent did he keep for himself?

POSTTEST

1. BASE is always_________ of a number.
2. Percent always has a_______ _______ .
3. Find the following percents:

   a) 13 is_____% of 65?
   b) 38 is_____% of 114?
   c) 16 is_____% of 96?
   d) 125 is____% of 100?

OBJECTIVE 6-94

FINDING 100% OF A NUMBER WHEN A PART OF IT IS GIVEN

OBJECTIVE: The student will be able to use a proportion to find (100%) the base of a number when a part of the number is known.

PRETEST

1. a) 14 is 10% of what number?
   b) 12 is 6% of what number?
   c) 5 is what percent of 10?
   d) 10 is 80% of what number?
   e) 2 is what percent of 20?
2. Set up a proportion to illustrate how to find 100% of a number when you know a part of it.

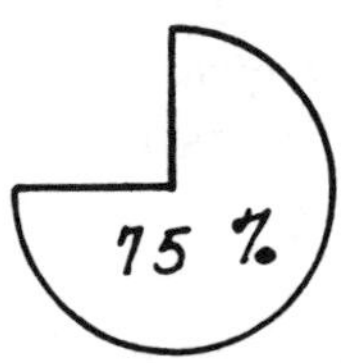

Compare these two circles. You can see tha the first circle is complete and the secon circle is not. The first circle is a whole circl It is 100% of a circle: it contains 360 degree The second circle is not a whole circle. Part c it is missing. What percent is missing in th second circle? How many degrees are missing 75% = _____degrees?
NOW, YOU ARE GOING TO LEARN HOW TO FIND 100% OF A NUMBER. WE WIL USE THE PROBLEM WITH THE CIRCLES

You have already figured out that there are 270 degrees in 75% (or 3/4) of a circle. This is how you would find the number of degrees in a whole circle or 100% of a circle: Make a ratio of the percent: $\frac{75}{100}$

We will let N = the number of degrees in 100% of a circle, because that is what we do not know and what we are trying to find. This time, the N will go in the denominator of the fraction, because it corresponds to the 100 in the fraction for the percent: $\frac{270}{N}$

75/100 = 270/N
Multiply 75 x N = 75N
Multiply 100 x 270 = 27,000

We now have this equation:

75N = 27,000 (Divide both sides by 75)
N = 360

Answer: There are 360 degrees in 100% of a circle.

Remember:

When you are looking for 100% of a number, always put the Unknown (N or some other symbol) in the denominator.

ANOTHER EXAMPLE:

Your friend had a test and received a mark of 85%. He had 17 examples correct. How many examples were on the test?

Solution: Make a ratio of the percent: $\frac{85}{100}$

Make a ratio of the number of examples correct to the number of examples on the test: $\frac{17}{N}$

Write an equation: $\frac{85}{100} = \frac{17}{N}$

Multiply 85 x N = 85N
Multiply 100 x 17 = 1700
85N = 1700 (Divide both sides by 85)
N = 20

Answer: There were 20 examples on the test.

STUDY THE ABOVE TWO EXAMPLES UNTIL YOU REALLY UNDERSTAND HOW TO DO THEM. THEN, GO ON TO THE PROBLEMS.

EXERCISE:

1. 10% of the students at McKay Campus School were absent on Thursday. 95 students were absent. How many students attend Mckay? (Be sure to set up proportions to solve these problems.)
2. Your friend bought some used comic books at a big discount. He gave you 10 of them which was 20% of the number he bought. How many used comics did he buy?
3. Your mother sent you to the store to buy some eggs. On the way home, you dropped the bag and broke 25% of the eggs. You broke 6 eggs. How many eggs did you buy?
4. You went to a movie with your sister. She made you pay 1/3 of the admission price out of your allowance. If you paid $.50, what was the admission price?
5. Betty spent $9.00 on a blouse. This was 200% of her weekly allowance. She did not have enough money, so her mother gave her the rest. What was Betty's weekly allowance?

POSTTEST

1. Solve these equations:

   a) 42/100 = N/50 b) 65/100 = 65/N c) 35/100 = N/20

2. Give equivalent fractions in lowest terms for these percents:

   a) 20% b) 25% c) 33 1/3% d)85% e) 125% f)10%
   g) 75% h) 50% i) 80%

3. a) 2 is 10% of______?
   b) 25 is 50% of______?
   c) 18 is 25% of______?
   d) 125 is 25% of_____?
   e) 200 is 5% of______?

OBJECTIVE 6-95

USING PROPORTION TO SOLVE PERCENT PROBLEMS

Prerequisite for this packet: Obj. 6-(91-94)

OBJECTIVE:

The student will be able to use proportion to solve problems involving the three types of percent:

1. Finding a percent of a number.
2. Finding what percent one number is of another.
3. Finding base (100%) of a number when a part of the number is given.

PRETEST

1. How many ratios are in a proportion?
2. Is an equation a proportion?
3. Is an inequality a proportion?
4. What does "percent" mean?
5. What does "base" mean in percent?

In the four packets you have done on ratio and percent, you have learned how to solve the three types of percent problems stated in the objective above. When you worked with these problems, you did one type at a time. It was not necessary for you to figure out what proportion you would use for each type, because you were shown. In this packet, you will have to figure out how to set up the proportion yourself.

FACTS TO REMEMBER:

1. When a percent is given, set up the percent as a ratio with a denominator of 100.
2. If a percent is not given, make a ratio of the two numbers that are given. Express the percent that you are trying to find by N/100.
3. If you are looking for base (100%) of a number, make a ratio of the percent. Put the number that is a part of the whole over N.
4. If you are looking for a percent of a number, make a ratio of the number you do not know to the number that is given. The N will be in the numerator.

The hard part of solving these problems will be to decide what you are asked to find in the problem. What is missing? Is it a part of a number, all of a number, or a number with a percent sign?

EXAMPLES OF HOW TO SET UP THE PROPORTIONS TO SOLVE EACH TYPE OF PERCENT PROBLEM:

To find a PART OF A NUMBER: Find 25% of 24

$$\frac{25}{100} = \frac{N}{24}$$

To find WHAT PERCENT ONE NUMBER IS OF ANOTHER:
What percent of 24 is 6?

$$\frac{N}{100} = \frac{6}{24}$$

To find 100% OF A NUMBER WHEN A PART OF THE NUMBER IS GIVEN AND A PERCENT TO REPRESENT THAT PART OF THE WHOLE NUMBER IS ALSO GIVEN:
6 is 25% of what number?

$$\frac{25}{100} = \frac{6}{N}$$

STUDY THESE EXAMPLES UNTIL YOU KNOW EXACTLY WHAT YOU ARE LOOKING FOR IN EACH TYPE OF PERCENT PROBLEM. THEN, WHEN YOU FEEL YOU ARE READY, GO ON TO THE POSTTEST.

POSTTEST

SET UP A PROPORTION FOR EACH EXAMPLE; THEN SOLVE FOR THE ANSWER:

1. Find the part of the whole that is asked for:

   a) 25% of 176 is________ .
   b) 16% of 1000 is_______.
   c) 33 1/3% of 75 is______.
   d) 3/4 of 128 is______.
   e) 3/8 of 72 is________.

(Change the fractions in d and e to equivalent fractions with denominators of 100.)

2. Find the percent:

a) 35 is ___% of 70
b) 33 is ___% of 165
c) 8 is ___% of 80
d) 9 is ___% of 12
e) 72 is ___% of 80

3. Find the base when a part of the number and a percent are given:

a) 16 is 25% of ____.
b) 19 is 75% of ____.
c) 100 is 100% of ____.
d) 13 is 10% of ____.
e) 45 is 15% of ____.

4. If you save $1.00 per week and that is 33 1/3% of your weekl allowance, how much is your weekly allowance?
5. If there are 180 days of school and you are absent 10% of the time, how many days are you present?
6. In the above problem, what percent of the time are you present?
7. In your class, there are three girls and four boys who wear glasses. If there are 28 pupils in the class, what percent of the class wear glasses?
8. If you got a mark of 75% on a spelling test and you had 18 words spelled correctly, how many words were there on the test?
9. If 75% of your closest friends own ten-speed bikes and they have 6 bikes among them, how many close friends do you ha
10. If 20% of the books you read are mysteries and you read abo 30 books a year, how many mystery stories do you read?

OBJECTIVES FOR LEVEL 7

The student will be able to:

7-1 Read and write numerals up to the quadrillions place.
7-2 Use the Commutative Property of Addition with integers.
7-3 Use the Commutative Property of Multiplication with integers.
7-4 Use the Associative Property of Addition with integers.
7-5 Use the Associative Property of Multiplication with integers.
7-6 Use the Distributive Property with integers.
7-7 Use the Property of Zero under Addition with integers.
7-8 Use the Property of Zero under Multiplication with integers.
7-9 Use the Property of Zero under Division with integers.
7-10 Use the Property of One under Multiplication with integers.
7-11 Use the Property of One under Division with integers.
7-12 Use the Special Property of a Number and Itself with integers.
7-13 Use the Property of Closure with integers.
7-14 Discuss some of the uses and identify the symbols for Roman Numerals: I, V, X, L, C, D, M.
7-15 Write given decimal numerals in Roman Numerals.
7-16 Identify terms used in the four basic operations: Addend, sum, minuend, subtrahend, difference, factor, multiplicand, multiplier, product, dividend, divisor, quotient, remainder.
7-17 Express fractions in lowest terms.
7-18 Find L.C.D. (Least Common Denominator) by factoring.
7-19 Add fractions with unlike denominators.
7-20 Add mixed numbers and fractions with unlike denominators.
7-21 Subtract fractions from whole numbers.
7-22 Subtract fractions from mixed numbers.
7-23 Subtract mixed numbers from whole numbers.
7-24 Multiply a fraction by a fraction.
7-25 Multiply a fraction by a whole number.
7-26 Multiply a fraction by a mixed number.
7-27 Multiply a mixed number by a whole number.
7-28 Multiply a mixed number by a mixed number.
7-29 Divide a whole number by a fraction.
7-30 Divide a fraction by a fraction.

7-31 Divide a mixed number by a whole number.
7-32 Divide a mixed number by a fraction.
7-33 Divide a mixed number by a mixed number.
7-34 Solve problems involving the four basic operations with fractions.
7-35 Add decimal fractions up to the hundred-thousandths place.
7-36 Subtract decimal fractions up to the hundred-thousandths place.
7-37 Multiply decimal fractions up to the hundred-thousandths place.
7-38 Divide decimal fractions up to the hundred-thousandths place.
7-39 Solve problems involving the four basic operations with decimal fractions.
7-40 Find decimal equivalents for fractions.
7-41 Find fractional equivalents for decimal fractions.
7-42 Change decimal fractions to percents.
7-43 Change percents to decimal fractions.
7-44 Use decimal fractions to find a percent of a number.
7-45 Divide to find what percent one number is of another.
7-46 Divide to find 100% (base) of a number when a part of the number is given.
7-47 Round off decimal fractions to a given place.
7-48 Identify repeating and terminating decimal fractions.
7-49 Find discount and net price.
7-50 Find regular price when sale price and rate of discount are given.
7-51 Find commission.
7-52 Find rate of commission.
7-53 Compute simple interest.
7-54 Find percent of increase and decrease.
7-55 Identify: Margin, overhead, profit, loss.
7-56 Solve problems involving profit and loss.
7-57 Make out deposit slips and write checks.
7-58 Make a budget and keep an account.
7-59 Compute compound interest.
7-60 Read an electric meter.
7-61 Check an electric light bill.

7-62 Read a gas meter.
7-63 Check a gas bill.
7-64 Compute the cost of buying given items on the installment plan.
7-65 Graph rational numbers.
7-66 Order a set of rational numbers.
7-67 Write reciprocals of given numbers.
7-68 Represent numbers in as many ways as possible (1/4, 1:4, 1 to 4, etc.).
7-69 Use the Sieve of Eratosthenes to find prime numbers up to 100.
7-70 Identify given subsets of the Whole Number System.
7-71 Make a scale drawing, computing all necessary data.
7-72 Read an automobile map.
7-73 Add integers.
7-74 Subtract integers.
7-75 Multiply integers.
7-76 Divide integers.
7-77 Construct and interpret circle graphs.
7-78 Find perimeter and circumference, using formulas for given figures.
7-79 Find area, using formulas for given figures.
7-80 Find volume, using formulas for given prisms.
7-81 Solve problems involving perimeter, circumference, area and volume.
7-82 Identify: Angles (right, acute, obtuse, straight, supplementary, complementary).
7-83 Use a protractor to measure given angles and state what kind of angles they are.
7-84 Use a protractor to construct angles to a given number of degrees.
7-85 Measure the angles in given triangles.
7-86 Identify: Scalene, isosceles, equilateral triangles.
7-87 Construct the above types of triangles.
7-88 Identify: Transversal, vertical angles, congruent angles, interior angles, exterior angles, adjacent angles.
7-89 Identify given quadrilaterals and state the properties of each.

7-90 Use the compass to create original designs.
7-91 Construct inscribed and circumscribed polygons.
7-92 Construct given polyhedra.
7-93 Use exponents to perform simple multiplication ($2^3 \times 3^2$).
7-94 State the place value of numerals in Base 5 up to the sixth place.
7-95 Change numerals from Base 5 to Base 10.
7-96 Change numerals from Base 10 to Base 5.
7-97 Write numerals up to 50 in Base 5.
7-98 Add and subtract in Base 5.
7-99 Multiply and divide in Base 5.
7-100 Define the prefixes used in the Metric System.
7-101 Express metric measures as powers of 10.
7-102 Use linear measure in the Metric System.
7-103 Solve problems involving linear measure in the Metric System.
7-104 Use equations to solve problems.

OBJECTIVE 7-1
READING AND WRITING NUMERALS

OBJECTIVE: The student will be able to read and write numerals up to the quadrillions place.

PRETEST

Read the following numerals to your teacher:

1) 79 2) 7,959 3) 674,364 4) 6,004,009
5) 2,876,593,417,362,051 6) 52,825,463,928,000

Write the following numerals as word statements:
7) 589 8) 9,207,800,000,000 9) 843,297,861

10) Write the numeral which names the following number:
Two billion, fifty-eight million, seven hundred sixty-seven thousand, five hundred ninety-three.

To read numerals containing 4 or more digits, first separate by commas the given numeral into as many groups of 3 digits (called periods) each as possible, starting from the units (ones) place and going to the left.

Next, starting from the left and going to the right, read each group of figures separately and read each comma properly as it is reached.

The first comma from the right is read "thousand," the second comma "million," the third "billion," the fourth "trillion" and the fifth "quadrillion."

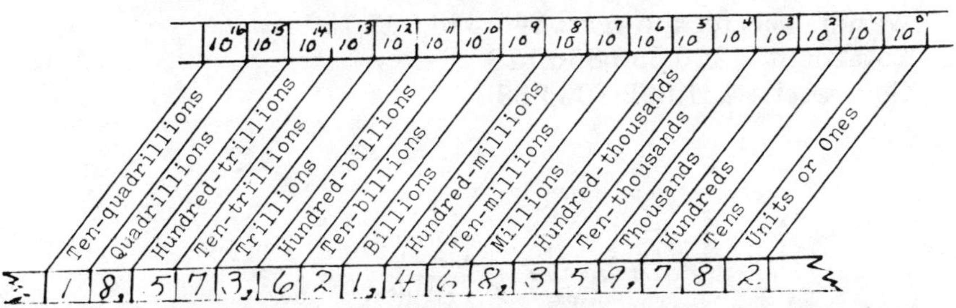

The above numeral is read and written: Eighteen quadrillion, five hundred seventy-three trillion, six hundred twenty-one billion, four hundred sixty-eight million, three hundred fifty-nine thousand, seven hundred eighty-two.

EXAMPLE:

1) Read and write as a word statement: 5,362,899,347
Answer: Five billion, three hundred sixty-two million, eight hundred ninety-nine thousand, three hundred forty-seven.

2) Read and write as a word statement: 2,540,250,000
Answer: Two billion, five hundred forty million, two hundred fifty thousand.

NOTICE THAT WE USE COMMAS TO SEPARATE AND NAME THE PERIODS WHEN WRITING NUMERALS.

EXAMPLE:

3) Write as a numeral: Thirty-four million, eighty thousand, two hundred ninety.
Answer: 34,080,290

EXERCISE: Read and write the following numerals given in these facts:

A U.S. Treasury report shows its cash position is:
Balance: $3,176,689,943
Deposits for the year: $33,789,916,198
Withdrawals for the year: $41,378,897,750
Total debt: $280,350,866,129
Gold assets: $21,692,706,238

POSTTEST

Read the following numerals to your teacher:

1) 84 2) 4,206,000 3) 900,281 4) 6,184,673,290,000,502
5) 7,281,063,305,664,763

Write the following numerals as word statements:
6) 68 7) 4,975 , 8) 3,582,942 9) 7,800,152,009,000,000

10) The Pacific Ocean has an area of sixty-three million, eight hundred one thousand, six hundred sixty-eight square miles. Write the area of the Pacific Ocean as a numeral.
11) The Atlantic Ocean has an area of thirty-one million, eight hundred thirty thousand, seven hundred eighteen square miles. Write the area of the Atlantic Ocean as a numeral.
12) Write as a word statement the difference in area of the two oceans above.
13) A one followed by one hundred zeros is called a "googol." How many periods does it have?

OBJECTIVES 7-(14,15)
ROMAN NUMERALS

OBJECTIVES:

The student will be able to:
Discuss some of the uses and identify the symbols for Roman Numerals: I, V, X,L, C, D, M.

Write given decimal numerals in Roman Numerals.

PRETEST

1. What uses can you think of for Roman Numerals? List as many as you can.
2. What decimal numeral does each of the following Roman Numerals represent?
a) V b) L c) D d) MM e) CD f) XC
g) LXVI h) XCVIII i) MDCCLXVI J) MMMCCCXXXIII

3. Write the Roman Numeral for each of the following numbers:
a) 7 b) 4 c) 19 d) 34 e) 38 f) 63
g) 76 h) 299 i) 459 j) 1970

In mathematics, the influence of the Romans was not very great. The Roman Numerals were the most popular contribution of the Romans and lasted until the middle of the 15th century.

Some of the most familiar uses for Roman numerals are in outlines, some clock dials, prefaces of books, book and chapter numbers and dates.

Roman numerals were formed by writing from left to right as a sum, first the symbol for the largest possible value, then the symbol for the next smaller value, etc.

The Romans used symbols for numbers:

ROMAN NUMERAL	I	V	X	L	C	D	M
DECIMAL NUMERAL	1	5	10	50	100	500	1000

The symbols I, X, C, and M may be used as many as three times when necessary. When the symbols I, X, or C precede (come before) a Roman number symbol of greater value, its value is subtracted from the larger value.

THE SYMBOLS V, L, AND D NEVER PRECEDE A SYMBOL OF GREATER VALUE and are NEVER USED IN SUCCESSION (one right after the other).

EXAMPLE 1: MDCCXCIII is what number?

Since

$$\begin{aligned} M &= 1000 \\ D &= 500 \\ CC &= 200 \\ XC = 100 - 10 &= 90 \\ III &= 3 \\ \hline MDCCXCIII &= 1793 \end{aligned}$$

EXAMPLE 2: Change 1995 to a Roman number.

$$\begin{array}{rr} M = & 1000 \\ CM = 1000 - 100 = & 900 \\ XC = 100 - 10 = & 90 \\ V = & 5 \\ \hline & 1995 = MCMXCV \end{array}$$

EXAMPLES:

IV = 5 − 1 = 4; XL = 50 − 10 = 40;
XXIX = 20 + (10 − 1) = 29;
89 = LXXXIX since 50 + 30 + (10 − 1) is equal to 89;
MMMCD = 3000 + (500 − 100) = 3400;
LXXII = 50 + 20 + 2 = 72

EXERCISES:

1) What number does each of the following Roman numerals represent?

a) IV b) LX c) CCCIX d) MLV
e) MMCDXLIV f) CMXCIX

2. Write the Roman numeral for each of the following numbers:
a) 6 b) 9 c) 12 d) 19 e) 29
f) 78 g) 666 h) 150 i) 804 j) 49

POSTTEST

1. What other uses for Roman Numerals have you thought of?
2. Do you see any advantages or disadvantages of Roman Numerals as opposed to decimal numerals?
3. What decimal number does each of the following Roman Numerals represent?

a) VII b) XIII c) XIV d) XXXIX e) XLIV
f) LXXVII g) CDXLI h) MCMLXV i) MMMDXCIX
j) MCMLXXVI

4. Write the Roman Numeral for each of the following numbers:
   a) 14 b) 17 c) 42 d) 89 e) 98 f)212
   g) 555 h) 999 i) 1995 j) 2798
5. Express the following Roman Numerals correctly:
   a) VVV b) VIIII c) LC d) DD e) CLL f) VL

OBJECTIVES 7-(18-20)
ADDITION OF FRACTIONS AND MIXED NUMBERS

<u>OBJECTIVES</u>: The student will be able to:
Find the Least Common Denominator (L.C.D.) by factoring.
Add fractions with <u>unlike</u> denominators.
Add mixed numbers and fractions with <u>unlike</u> denominators.

<u>PRETEST</u>

1) Find <u>all</u> the factors of these numbers:
   a) 14 b) 12 c) 6 d) 18 e) 32
2) Find the L.C.D. of the following:
   a) 1/2; 2/3 b) 1/6; 2/3 c)1/4; 1/6 d) 1/8; 1/3
   e) 1/4; 1/12

3) Add:

a) 1/2 + 2/3 =
b) 1/6 + 2/3 =
c) 1/4 + 1/6 =
d) 1/8 + 1/3 =
e) 1/4 + 1/2 =

4) Add:

a) 1/2 + 2 1/4 =
b) 1/4 + 3 1/3 =
c) 1/8 + 6 2/3 =
d) 1/5 + 2 1/4 =
e) 2/5 + 1 1/6 =

When adding fractions, you must first find the L.C.D. (Least or Lowest Common Denominator). To find the L.C.D., you must break the numbers down into their prime factors. Then, cross out the common factors in both factorizations. After crossing out the common factors, multiply the numbers by the uncommon factors of the other numbers.

EXAMPLE: Find the L.C.D. of 1/4 and 1/3

Step 1: Find the prime factorizations of 3 and 4;
3 is a prime number, so its factors are 3 and 1
4 = 2 x 2, so they have no common factors.

Step 2: Multiply the two numbers. The L.C.D. is 2 x 2 x 3 or 12

EXAMPLE: Find the L.C.D. of 1/6 and 1/8

Step 1: Find the prime factorizations of 6 and 8
$6 = 3 \times 2$
$8 = 2 \times 2 \times 2$

Step 2: Cross out a 2 in each factorization, since 2 is a common factor. This leaves 3 and 2 x 2.

Step 3: Multiply 6 by 2 x 2 and multiply 8 by 3.
The L.C.D. is 24

To add two or more fractions with unlike denominators, first find the L.C.D. Change the fractions to equivalent fractions with the L.C.D. as the denominator. Add the numerators and put the result over the L.C.D. Reduce to lowest terms, if necessary.

EXAMPLE:

$$\frac{1}{3} + \frac{1}{5}$$

Step 1: The L.C.D. is 15. Change each fraction to an equivalent fraction with 15 as the denominator:

$$\frac{1}{3} = \frac{5 \times 1}{5 \times 3} = \frac{5}{15} \text{ AND } \frac{1}{5} = \frac{3 \times 1}{3 \times 5} = \frac{3}{15}$$

So, $\frac{1}{3} + \frac{1}{5} = \frac{5}{15} + \frac{3}{15}$

Step 2: $\frac{5}{15} + \frac{3}{15} = \frac{8}{15}$

so, $\frac{1}{3} + \frac{1}{5} = \frac{8}{15}$ (Answer)

EXAMPLE: $\frac{2}{3} + \frac{1}{6}$

Step 1: The L.C.D. is 6. Change each fraction to an equivalent fraction with 6 as the denominator.

$$\frac{2}{3} = \frac{2 \times 2}{2 \times 3} = \frac{4}{6}$$

Step 2: $\frac{4}{6} + \frac{1}{6} = \frac{5}{6}$

so, $\frac{2}{3} + \frac{1}{6} = \frac{5}{6}$ (Answer)

When adding a mixed number and a fraction, change the mixed number to an improper fraction. Then, proceed as if adding two fractions.

EXAMPLE: 2 1/4 + 1/5

Step 1:

Change 2 1/4 to an improper fraction by multiplying the whole number (2) by the denominator (4) and adding the numerator (1). Put this number over the denominator (4).

2 x 4 = 8. Add 1.
8 + 1 = 9. Put 9 over the denominator 4:

2 1/4 = 9/4

Step 2: $\frac{9}{4} + \frac{1}{5}$ The L.C.D. is 20

STEP 3: $\frac{9}{4} = \frac{5 \times 9}{5 \times 4} = \frac{45}{20}$ AND $\frac{1}{5} = \frac{4 \times 1}{4 \times 5} = \frac{4}{20}$

Step 4: $\frac{45}{20} + \frac{4}{20} = \frac{49}{20}$

Step 5: $\frac{49}{20}$ = 2 9/20

so, 2 1/4 + 1/5 = 2 9/20 (Answer)

EXERCISES

1) Find the L.C.D.:
a) 1/4; 1/3 b) 1/8;1/2 c) 1/7; 1/14 d) 1/4; 1/12
e) 1/6; 1/3
2) Add the following:
a) 1/4 + 1/3 b) 1/8 + 1/2 c) 1/7 + 1/14
3) Change to improper fractions:
a) 2 1/3 b) 3 3/4 c) 1 1/6 d) 5 1/8 e) 2 1/5
4) Add the following:
a) 1/4 + 2 1/3 b) 2 1/8 + 1/4 c) 1/4 + 2 1/6

POSTTEST

1) Find the L.C.D.:
a) 1/4;2/3 b) 1/8;1/6 c) 1/6;3/4 d) 1/7;1/3
e) 2/3;1/5
2) Add the following:
a) 2/3 + 1/8 b) 1/6 + 4/7 c) 1/8 + 3/4
d) 7/8 + 1/24 e) 5/6 + 1/8
3) Add the following:
a) 2 1/3 + 4/7 b) 3 1/4 + 1/3 c) 1 1/7 + 1/4
d) 4 1/6 + 1/8 e) 6 1/9 + 1/3

(OBJECTIVES 7-(21-23)
SUBTRACTION OF FRACTIONS AND MIXED NUMBERS

OBJECTIVES: The student will be able to:
Subtract fractions from whole numbers.
Subtract fractions from mixed numbers.
Subtract mixed numbers from whole numbers.

PRETEST

1) SUBTRACT:
   a) 4 − 3/4 = b) 7 − 1/2 = c) 3 − 1/3 =
   d) 5 − 1/6 = e) 8 − 1/5 =

2) SUBTRACT:
   a) 2 1/3 − 2/3 = b) 4 1/4 − 1/2 = c) 5 3/7 − 1/7 =
   d) 6 3/8 − 1/2 = e) 7 2/3 − 1/6 =

3) SUBTRACT:
   a) 7 − 2 1/3 = b) 8 − 6 1/2 = c) 4 − 2 1/8 =
   d) 4 − 1 2/3 = e) 10 − 5 1/5 =

When subtracting fractions from whole numbers, you must first change the whole number to an improper fraction. Then you must find the Least Common Denominator (L.C.D.). That is, you must find the smallest number that can be divided evenly by both denominators.

EXAMPLE: 5 − 1/4

Step 1: Change 5 to an improper fraction: 5 = 5/1

Step 2: Find the L.C.D.

$$\frac{5}{1} = \frac{4 \times 5}{4 \times 1} = \frac{20}{4} \quad \text{(The L.C.D. is 4)}$$

Step 3: Subtract:

$$\frac{20}{4} - \frac{1}{4} = \frac{19}{4} \quad \text{(Subtract numerator from numerator)}$$

$$\frac{19}{4} = 4\,\frac{3}{4}$$

So, 5 − 1/4 = 4 3/4 (Answer)

EXERCISE:

SUBTRACT:
1) 10 − 1/5 =
2) 25 − 1/6 =
3) 4 − 1/3 =

When subtracting fractions from mixed numbers, change the mixed number to an improper fraction. Find the L.C.D. and subtract. Then, express the answer in the form of a mixed number.

EXAMPLE: 7 1/2 – 3/4

Step 1: Change the mixed number to an improper fraction:

$7\ 1/2 = \frac{15}{2}$

$\frac{15}{2} - \frac{3}{4}$ (The L.C.D. is 4)

Step 2: Change $\frac{15}{2}$ to an equivalent fraction having a denominator of 4:

$\frac{15}{2} = \frac{2 \times 15}{2 \times 2} = \frac{30}{4}$

Step 3: Subtract:

$\frac{30}{4} - \frac{3}{4} = \frac{27}{4}$

$\frac{27}{4} = 6\ 3/4$

So, 7 1/2 – 3/4 = 6 3/4 (Answer)

When subtracting a mixed number from a whole number, change both numbers to improper fractions. Find the L.C.D. and subtract. Change the answer to a mixed number.

EXAMPLE: 7 – 2 1/3

Step 1: Change both numbers to improper fractions:

$7 = \frac{7}{1}$ AND $2\ 1/3 = \frac{7}{3}$

$\frac{7}{1} - \frac{7}{3}$ (The L.C.D. is 3)

Step 2: Change $\frac{7}{1}$ to an equivalent fraction having a denominator of 3:

$$\frac{7}{1} = \frac{3 \times 7}{3 \times 1} = \frac{21}{3}$$

Step 3: Subtract:

$$\frac{21}{3} - \frac{7}{3} = \frac{14}{3}$$

$$\frac{14}{3} = 4\ 2/3$$

So, 7 − 2 1/3 = 4 2/3 (Answer)

EXERCISE:

1) 2 1/4 − 1/2 =
2) 10 4/7 − 6/7 =
3) 6 1/4 − 5/8 =
4) 7 − 2 1/3 =
5) 24 − 6 1/4 =
6) 18 − 4 1/6 =

POSTTEST

1) SUBTRACT:

a) 12 − 1/9 =
b) 17 − 8/11 =
c) 9 − 4/5 =
d) 11 − 3/8 =
e) 15 − 1/8 =

2) SUBTRACT:

a) 2 5/12 − 1/6 =
b) 6 2/3 − 1/8 =
c) 4 6/7 − 3/4 =
d) 14 8/9 − 17/18 =
e) 7 2/7 − 6/7 =

3) SUBTRACT:

a) 14 − 2 1/7 =
b) 17 − 7 8/11 =
c) 28 − 2 4/5 =
d) 13 − 6 1/4 =
e) 15 − 10 3/8 =

OBJECTIVES 7 - (24-28)
MULTIPLICATION OF FRACTIONS AND MIXED NUMBERS

<u>OBJECTIVES:</u> The student will be able to:

Multiply a fraction by a fraction.
Multiply a fraction by a whole number.
Multiply a fraction by a mixed number.
Multiply a mixed number by a whole number.
Multiply a mixed number by a mixed number.

PRETEST

1) Multiply:
   a) 2/3 x 1/4 b) 4/7 x 2/3 c) 1/8 x 1/4
2) Multiply:
   a) 1/2 x 7 b) 2/3 x 6 c) 1/8 x 8

3) Multiply:
   a) 1/2 x 2 1/4   b) 4/5 x 2 2/5   c) 2/3 x 2 1/4
4) Multiply:
   a) 2 1/4 x 4   b) 7 1/4 x 2   c) 3 1/4 x 7
5) Multiply:
   a) 6 1/4 x 2 1/7   b) 4 2/7 x 1 1/3   c) 4 1/8 x 2 2/3

When multiplying a fraction by a fraction, multiply the numerators of the fractions together to get the numerator of the product. Multiply the denominators of the fractions together to get the denominator of the product. Express the fraction in lowest terms.

EXAMPLE:

1/4 x 2/3

Step 1: Multiply numerator by numerator and denominator by denominator:

$$1/4 \times 2/3 = \frac{1 \times 2}{4 \times 3} = \frac{2}{12} = \frac{1}{6}$$

So, 1/4 x 2/3 = 1/6 (Answer)

When multiplying a fraction by a whole number, change the whole number to an improper fraction. Then, multiply numerator by numerator and denominator by denominator. Reduce to lowest terms and change the answer to a mixed number.

EXAMPLE: 3/4 x 6

Step 1: Change 6 to an improper fraction: $6 = \frac{6}{1}$

Step 2: Multiply numerator by numerator and denominator by denominator:

$$\frac{3 \times 6}{4 \times 1} = \frac{18}{4}$$

Step 3: Reduce, if possible: $\frac{18}{4} = \frac{9}{2}$

Step 4: Change to a mixed number: $\frac{9}{2} = 4\ 1/2$

So, 3/4 x 6 = 4 1/2 (Answer)

When multiplying a fraction by a mixed number, change the mixed number to an improper fraction. Then, multiply numerator by numerator and denominator by denominator and change the answer to a mixed number.

EXAMPLE: 3/7 x 2 1/4

Step 1: Change 2 1/4 to an improper fraction: 2 1/4 = $\frac{9}{4}$

Step 2: Multiply numerator by numerator and denominator by denominator:

$$\frac{3}{7} \times \frac{9}{4} = \frac{3 \times 9}{7 \times 4} = \frac{27}{28} \qquad \text{(in lowest terms)}$$

so, 3/7 x 2 1/4 = $\frac{27}{28}$ (Answer)

When multiplying a mixed number by a whole number, change both the mixed number and the whole number to improper fractions. Then, multiply as above.

EXAMPLE: 2 3/4 x 6

Step 1: Change both numbers to improper fractions:

$2\ 3/4 = \frac{11}{4}$ AND $6 = \frac{6}{1}$

Step 2: Multiply: $\frac{11}{4} \times \frac{6}{1} = \frac{11 \times 6}{4 \times 1} = \frac{66}{4}$

Step 3: Reduce: $\frac{66}{4} = \frac{33}{2}$

Step 4: Change to a mixed number: $\frac{33}{2} = 16\ 1/2$

so, 2 3/4 x 6 = 16 1/2

When multiplying a mixed number by a mixed number, change both mixed numbers to improper fractions. Then, multiply.

EXAMPLE:

6 1/5 x 2 2/3

Step 1: Change to improper fractions:

$6\ 1/5 = \frac{31}{5}$ AND $2\ 2/3 = \frac{8}{3}$

Step 2: Multiply:

$$\frac{31}{5} \times \frac{8}{3} = \frac{31 \times 8}{5 \times 3} = \frac{248}{15}$$

Step 3: Reduce: $\frac{248}{15}$

(No common factors, so cannot be reduced. Divide 15 into 248 and put the remainder over 15.)

248 divided by 15 is 16 8/15

so, 6 1/5 x 2 2/3 = 16 8/15 (Answer)

EXERCISES:

a) 1/4 x 3/4 b) 2/3 x 1/7 c) 3/7 x 6
d) 3/4 x 7 e) 2/3 x 3 1/4 f) 5/8 x 3 1/2
g) 4 3/4 x 7 h) 2 3/5 x 10 i) 3 2/3 x 2 1/4
j) 6 2/5 x 7 1/8

POSTTEST

1) Multiply:
a) 1/4 x 1/5 b) 2/5 x 2/3 c) 3/4 x 3/8
d) 1/8 x 2/9 e) 3/5 x 3/8

2) Multiply:

a) 2/3 x 6   b) 1/4 x 8   c) 3/5 x 15
d) 7/8 x 16   e) 2/7 x 9

3) Multiply:

a) 1/4 x 3 1/2   b) 1/8 x 6 2/3   c) 3/5 x 2 1/8
d) 2/3 x 3 7/8   e) 2/5 x 4 1/9

4) Multiply:

a) 2 1/4 x 7   b) 4 2/3 x 4   c) 2 1/4 x 6
d) 8 1/7 x 6   e) 7 1/4 x 8

5) Multiply:

a) 2 1/4 x 2 1/2   b) 7 1/8 x 4 1/5   c) 4 3/4 x 6 3/8
d) 6 1/4 x 2 3/4   e) 4 1/7 x 6 1/8

OBJECTIVES 7-(29-33)
DIVISION OF FRACTIONS AND MIXED NUMBERS

<u>OBJECTIVES</u>: The student will be able to:

Divide a whole number by a fraction.
Divide a fraction by a fraction.
Divide a mixed number by a whole number.
Divide a mixed number by a fraction.
Divide a mixed number by a mixed number.

<u>PRETEST</u>

1) <u>Divide</u>:
   a) 8 ÷ 1/2 b) 7 ÷ 1/3 c) 4 ÷ 2/3
2) <u>Divide</u>:
   a) 1/2 ÷ 2/3 b) 2/5 ÷ 1/2 c) 3/8 ÷ 1/4
3) <u>Divide</u>:
   a) 2 1/4 ÷ 4 b) 6 1/3 ÷ 3 c) 3 1/3 ÷ 10
4) <u>Divide</u>:
   a) 1 2/3 ÷ 1/2 b) 6 1/8 ÷ 2/3 c) 2 3/4 ÷ 1/7
5) <u>Divide</u>:
   a) 6 1/4 ÷ 2 1/3 b) 8 7/8 ÷ 1 3/4 c) 7 1/8 ÷ 2 1/4

When dividing a <u>whole number</u> by a <u>fraction,</u> change the whole number to an improper fraction. Use the <u>reciprocal</u> of the divisor (the number you are dividing by) and multiply.

<u>EXAMPLE:</u> 6 ÷ 1/2

<u>Step 1</u>: Change 6 to an improper fraction: $6 = \frac{6}{1}$

<u>Step 2</u>: Find the <u>reciprocal</u> of the divisor:
The <u>reciprocal</u> of 1/2 is $\frac{2}{1}$ (switch the numerator and denominator)

Step 3: Multiply:

$$\frac{6}{1} \times \frac{2}{1} = \frac{6 \times 2}{1 \times 1} = \frac{12}{1} = 12$$

NOTE: Notice that when a whole number is divided by a fraction, the quotient is larger than the number being divided.

So, 6 ÷ 1/2 = 12 (Answer)

When dividing a fraction by a fraction, find the reciprocal of the divisor and multiply.

EXAMPLE: 1/2 ÷ 1/4

Step 1: The reciprocal of 1/4 is $\frac{4}{1}$

Step 2: Multiply:

$$\frac{1}{2} \times \frac{4}{1} = \frac{1 \times 4}{2 \times 1} = \frac{4}{2} = 2$$

So, 1/2 ÷ 1/4 = 2 (Answer)

When dividing a mixed number by a whole number, change both the mixed number and the whole number to improper fractions. Find the reciprocal of the divisor and multiply.

EXAMPLE: 2 3/8 ÷ 3

Step 1: Change both numbers to improper fractions:

2 3/8 = $\frac{19}{8}$ AND 3 = $\frac{3}{1}$

Step 2: The reciprocal of $\frac{3}{1}$ is $\frac{1}{3}$

Step 3: Multiply:

$$\frac{19}{8} \times \frac{1}{3} = \frac{19 \times 1}{8 \times 3} = \frac{19}{24}$$ (In lowest terms)

So, 2 3/8 ÷ 3 = $\frac{19}{24}$ (Answer)

When dividing a mixed number by a fraction, change the mixed number to an improper fraction. Find the reciprocal of the divisor and multiply.

EXAMPLE: 4 1/8 ÷ 1/2

Step 1: Change 4 1/8 to an improper fraction:

$4\ 1/8 = \frac{33}{8}$

Step 2: The reciprocal of 1/2 is $\frac{2}{1}$

Step 3: Multiply:

$$\frac{33}{8} \times \frac{2}{1} = \frac{33 \times 2}{8 \times 1} = \frac{66}{8}$$

Step 4: Reduce: $\frac{66}{8} = \frac{33}{4}$

Step 5: Change to a mixed number:

$$\frac{33}{4} = 8\ 1/4$$

so, 4 1/8 ÷ 1/2 = 8 1/4 (Answer)

When dividing a mixed number by a mixed number, change both mixed numbers to improper fractions. Find the reciprocal of the divisor and multiply.

EXAMPLE: 2 3/4 ÷ 1 1/2

Step 1: Change both numbers to improper fractions:

$2\ 3/4 = \frac{11}{4}$ AND $1\ 1/2 = \frac{3}{2}$

Step 2: The reciprocal of $\frac{3}{2}$ is $\frac{2}{3}$

Step 3: Multiply:

$$\frac{11}{4} \times \frac{2}{3} = \frac{11 \times 2}{4 \times 3} = \frac{22}{12}$$

Step 4: Reduce:

$$\frac{22}{12} = \frac{11}{6}$$

Step 5: Change to a mixed number:

$\frac{11}{6} = 1\ 5/6$

So 2 3/4 ÷ 1 1/2 = 1 5/6 (Answer)

EXERCISES:

1) Find the reciprocals:
   a) 1/2 b) 3/4 c) 2/5 d) 3/8 e) 4/5
   First, change these mixed numbers to improper fractions, then find the reciprocals:
   f) 2 1/4 g) 6 2/3 h) 1 3/4 i) 7 2/5 j) 4 1/8
2) Divide:
   a) 6 ÷ 3/4 b) 2/3 ÷ 1/4 c) 4 1/8 ÷ 3
   d) 6 2/3 ÷ 2/5 e) 2 1/8 ÷ 2 1/4

POSTTEST

1) Divide:
   a) 4 ÷ 3/4 b) 7 ÷ 2/5 c) 14 ÷ 1/7 d) 25 ÷ 4/5
   e) 12 ÷ 2/3
2) Divide:
   a) 1/4 ÷ 2/7 b) 3/4 ÷ 1/8 c) 1/3 ÷ 1/5 d) 2/3 ÷ 7/8
   e) 4/5 ÷ 1/2
3) Divide:
   a) 4 1/6 ÷ 4 b) 4 1/8 ÷ 10 c) 6 1/2 ÷ 5
   d) 8 2/3 ÷ 3 e) 1 1/4 ÷ 8
4) Divide:
   a) 6 1/8 ÷ 3/4 b) 6 2/3 ÷ 1/4 c) 4 3/4 ÷ 4/5
   d) 4 1/7 ÷ 2/3 e) 5 1/5 ÷ 1/8
5) Divide:
   a) 10 1/8 ÷ 4 1/8 b) 6 2/3 ÷ 2 1/5 c) 8 7/8 ÷ 17 3/4
   d) 12 1/2 ÷ 6 1/4 e) 4 1/9 ÷ 2 1/3

OBJECTIVE 7-49
DISCOUNT AND NET PRICE

<u>OBJECTIVE:</u>

The student will be able to find the discount and net price of given items.

PRETEST

Find the following discounts and net prices:

1) SHOES: regular price, \$20; 10% discount.
2) SOFA: regular price, \$300; 20% discount.
3) COATS: regular price, \$20; 15% discount.
4) On a 20% discount sale, Mrs. Jones bought a chair originally marked \$40. What was the amount of the discount? What did she pay for the chair (net price)?

5) In a store, a group of furniture was marked "30% off." This means that a person who bought some of it would receive a 30% discount on the amount of the purchase. What would the amount of the discount be on a chair priced $60 before the sale?

When you go shopping in a store that has a sale on several items, you may see signs that read "All winter coats on sale at 30% discount" or "30% off the regular price of these winter coats." This means that if you buy a winter coat that regularly sold for $60, the amount of the discount would be 30% of $60. Thirty percent of $60 is $18 ($60 x .3). You would subtract $18 from $60 and would pay only $42 for the coat.

In the example above, 30% is the <u>rate</u> of discount; $60 is the <u>regular price</u> or the <u>base;</u> $18 is the <u>amount</u> of discount; and $42 is the <u>net price</u>.

<u>EXAMPLE 1:</u>

Mrs. Smith bought a table originally marked $45 on the 20% discount sale. What was the amount of the discount? What did she pay for the chair?

<u>Let</u>:	Amount of discount	=	?
	Base price	=	$45
	Rate of discount	=	20% = .2

<u>AMOUNT</u>	=	BASE X RATE
	=	$45 x 20%
	=	$45 x .2
	=	$9 <u>discount</u>
<u>NET PRICE</u>	=	BASE – AMOUNT OF DISCOUNT
	=	$45 – $9
	=	$36 net price

Mrs. Smith pays $36 for the table.

<u>EXAMPLE 2:</u>

John bought a new pair of skis from the neighborhood sporting goods store for $150. The next day, the newspaper advertised a 10% discount sale on the same pair of skis, regularly $150. How much more did John pay for the skis? What is the new price?

Let:		
	Amount of discount	= ?
	Base price	= $150
	Rate of discount	= 10% = .1
	AMOUNT	= BASE X RATE
		= 150 x .1
		= $15 (This is how much extra John paid.)
	NET PRICE	= BASE – AMOUNT
		= $150 – $15
		= $135 (This is the new price.)

EXERCISES:

If a store has a 15% discount sale on all items in the store, what is the amount of discount and the net price for each of the following items?

	ITEM	BASE (REGULAR PRICE)	AMOUNT OF DISCOUNT	NET PRICE
1)	Winter dresses	$20	?	?
2)	Sweater	$10	?	?
3)	Radio	$60	?	?
4)	Shoes	$18	?	?

5) What is the rate of discount at a half-price sale? What is the amount of discount on an article marked $9.50?

BE SURE TO STUDY THE EXAMPLES CAREFULLY. TAKE THE POSTTEST WHEN YOU FEEL YOU ARE READY.

POSTTEST

1) At an "after Christmas" sale, a store sold costume jewelry at a 10% discount. Beth bought a ring originally selling for $6.

What was the amount of discount? What was the net price?

2) At the same sale mentioned in Exercise 1, Ann bought a necklace that cost $5.50 before the sale. What was the amount of discount? What was the net price?
3) Bill works in a sporting goods store. All employees get a 15% discount on items they buy from the store. What was the amount of discount Bill received on the purchase of a tennis racket costing $32? How much did he pay for the tennis racket?
4) A television in a department store was marked "25% off." If the T.V. originally cost $480, how much was the amount of discount? What was the net price of the set?
5) All electric refrigerators were advertised at 15% discount on a month-end sale. How much discount would there be on a refrigerator originally marked $380?
6) Two stores, A-Mart and Bradlers, each have a sale on the same cassette recorder, regularly priced at $120. A-Mart has a "one-fourth off" sale and Bradlers has a "20% off" sale. Which store would you buy the recorder from and why? How much would you pay?

OBJECTIVES 7-(51,52)
COMMISSION

OBJECTIVES: The student will be able to:

Find commission.

Find rate of commission.

DEFINITION

Commission is a percentage of the selling price of an item which a salesman earns through the sale of that item.

PRETEST

1) A salesman receives commission of 6% for selling homes. He sold a house for $10,800. What was the amount of his commission?
2) An automobile salesman sold a car for $3200. He received a commission of 8% for selling the car. What was the amount of commission?
3) A salesman earned $2040 through the sale of a house worth $34,000. What was the <u>rate</u> of commission?
4) A T.V. salesman sold 5 televisions worth $300 each. His commission is 6% of all that he sells. How much commission did he earn by selling the 5 televisions?

Workers in many businesses are paid by the hour, week, month or year. They earn a given amount of money for each hour, week, etc. that they work.

Others, however, may work for commission. Salespeople who work for commission earn a percentage of each sale that they make. Men and women who sell homes, cars, furniture, appliances, TVs, stereos, etc., often work very hard when paid a commission, since the more they sell the more money they earn.

It is easy to figure out a salesperson's commission. We use the formula below:

AMOUNT OF COMMISSION = TOTAL SALES X RATE OF COMMISSION

EXAMPLE 1: A salesperson sells a home for $15,000. The commission is 6%. What is the amount of commission that the salesperson receives for the sale?

Amount of commission = ?
Total sales = $15,000
Rate of commission = $6\% = \frac{6}{100} = .06$

AMOUNT = TOTAL SALES X RATE
= $15,000 x 6%
= $\$15,000 \times \frac{6}{100}$

OR

= $15,000 x .06
= $900

The amount of commission the salesperson earns is $900.

EXAMPLE 2:

A car salesperson sells two cars, one for $3500 and the second for $4000. His commission is 4%. What is the amount of commission he earns?

Amount of commission = ?
Total sales = 7500 (3500 + 4000)
Rate of commission = 4% = $\frac{4}{100}$ = .04

AMOUNT = TOTAL SALES X RATE

$$= \$7500 \times \frac{4}{100}$$

OR

$$\$7500 \times .04$$

$$= \$300$$

The amount of commission the salesperson earned is $300.

EXERCISES:

1) What is the commission a salesperson receives when a house is sold for $33,000? The rate of commission is 5%.
2) A washing machine was sold for $250 and a dryer for $200. The commission was 4%. What was the amount of commission?
3) At a 5% commission, how much money would be earned on an automobile selling for $6,000?
4) A salesperson sold a receiver for $300, a turntable for $150, 2 speakers for $100 each and headphones for $50. What was the amount of the sale? If the commission was 3%, how much was earned from this sale?

BE SURE YOU UNDERSTAND HOW TO FIND AMOUNT OF COMMISSION BEFORE COMPLETING THE REST OF THIS PACKET.

It is sometimes necessary to find the rate of commission a salesperson earns when we know the amount of commission and the total sales. To find the rate, use this formula:

$$\text{RATE OF COMMISSION} = \frac{\text{AMOUNT OF COMMISSION}}{\text{TOTAL SALES}}$$

EXAMPLE 3:

$28 is earned on the sale of $700 worth of furniture. What is the rate of commission?

Rate of commission = ?
Amount of commission = $28
Total sales = $700

$$\text{RATE} = \frac{\text{AMOUNT}}{\text{TOTAL SALES}}$$

$$= \frac{28}{700} = \frac{4}{100} = .04 = 4\%$$

The rate of commission is 4%.

STUDY THESE EXAMPLES CAREFULLY BEFORE GOING ON TO THE EXERCISES.

EXERCISES:

1) A real-estate salesperson made $900 on the sale of a house valued at $18,000. What was the rate of commission?
2) Mr. Murphy decided to sell his home for $25,000. He could save $1000 by selling the house himself rather than letting the real-estate agent sell it. What was the rate of commission charged by the real-estate agent?

POSTTEST

1) An automobile sold for $2500. If the rate of commission was 8%, what was the amount of commission?
2) What was the commission on the sale of a building which sold for $150,000, if the rate of commission was 5%?
3) The Jones family listed their house with a real-estate

company. The company received 6% commission for selling the house. If the house sold for $14,500 how much was the commission? How much did the Jones family receive?

5) By selling his home without the involvement of a real-estate company, Mr. Smart saved $1540. If his home sold for $22,000, what rate of commission did the real-estate company charge?

6) Mrs. Brown made $1560 in one month by selling one home for $16,000 and a second for $10,000. What was her rate of commission?

OBJECTIVE 7-53
SIMPLE INTEREST

OBJECTIVE: The student will be able to compute simple interest.

DEFINITIONS:

INTEREST is money paid for the use of money.

PRINCIPAL is the money borrowed in a loan or invested (saved in a bank) on which interest is paid.

AMOUNT is the sum of the principal and the interest.

RATE is the percent of the principal paid as interest.

PRETEST

Using the formula i = prt where i = interest; p = principal; t = time; and r = rate, complete the following table:

INTEREST =	Principal	x Rate	x Time	
1.	?	$100	7% per yr.	1 year
2.	?	$100	7% per yr.	2 years
3.	?	$100	6% per yr.	6 months
4.	?	$400	2% per mo.	12 months
5.	?	$600	2% per mo.	$\frac{1}{4}$ year
6.	?	$500	3% per yr.	4 months

7) A man borrowed $600 at 4% simple interest for 3 years. How much interest did he owe? What was the amount due?

When you save money (invest) in a savings bank, the bank will pay you interest on this money (called the principal) for the time you keep your money in the bank.

For example, suppose a bank pays a rate of 6% simple interest per year, and you deposit (put money in a savings account) $100 principal. At the end of one year's time, you want to close your account (take out all of your money). The bank gives you your $100 back, plus $6 interest, since 6% of the $100 is $6 (.06 x 100 = 6). The amount the bank pays is the sum of the principal ($100) and the interest ($6); so amount = $106.

Let's try using the formula i = prt;
p = $100 principal
r = 6% per year (rate) $= \frac{6}{100}$ or .06

t= 1 year (time)

i = prt (principal x rate x time)

$i = 100 \times \frac{6}{100} \times 1$

OR

i= 100 x .06 x 1

i= \$6 (interest)   so, amount = \$100 + \$6 = \$106

EXAMPLE 2:

If you decided to save \$300 for 2 years at a bank paying 5% simple interest per year, use the formula i = prt to find the amount the bank would pay you.

i = prt

i = 300 x .05 x 2

i = \$30

so, the amount the bank pays you is \$330

EXAMPLE 3:

What is the amount due on a \$420 savings account at an annual (yearly) rate of 5% when a person takes all of the money out of the bank at the end of two years and six months?

p = \$420

$r = 5\% = \frac{5}{100}$ or .05

$t = 2\ 1/2$ years (2 years, 6 months $= 2\frac{6}{12} = 2\ 1/2$ years)

i = prt

i = 420 x .05 x 2 1/2

i = \$52.50 (interest)

so, amount= \$420 + 52.50 = \$472.50 (the amount the bank gives to the person)

We also compute simple interest on money borrowed from a bank. This is money you have to pay to the bank in addition to the principal.

EXAMPLE 4:

Bob borrowed (got a loan) $500 from a bank to buy a car. He agreed to pay back the loan after 1 year and six months, with the additional 8% interest rate per year.
How much simple interest did he pay? What was the amount he had to pay the bank?

$p = \$500$
$r = 8\% = \frac{8}{100}$ or .08
$t = 1\frac{6}{12} = 1\ 1/2$ years
$i = prt$
$i = 500 \times .08 \times 1\ 1/2$ or $500 \times .08 \times 1.5$
$i = \$60$
so, amount = $500 + 60 = $560 to be repaid.

STUDY THE EXAMPLES CAREFULLY BEFORE GOING ON TO THE EXERCISES.

EXERCISE:

Make believe that you are a bank manager. You have loaned out money to 5 people today. How much interest will your bank collect from each of these loans? Use i = prt. Change the time to years when necessary.

Name	Simple Interest =	Principal x	Rate x	Time
1. Bob	?	$100	4% per yr.	6 mos.
2. Sue	?	$200	8% per yr.	2 yrs.
3. Mary	?	$300	6% per yr.	8 mos.
4. Jeff	?	$400	6½% per yr.	1½ yrs.
5. Tim	?	$4500	4% per yr.	9 mos.

6) If you saved $50 in a bank paying 8% simple interest per year, how much interest do you get every six months?

7) Bert has $180 in a savings bank. The bank pays 6% interest per year. How much interest will Bert receive at the end of 6 months if he takes all his money out of the bank? What is the total amount Bert will receive?

POSTTEST

Complete the table. Use i = prt. Change time to the same units as the rate. (If r = 4% per month and t = 1 1/2 yrs., then 1 1/2 yrs. = 18 months.)

	Simple Interest	Principal	Rate	Time
1.	?	$900	4% per yr.	¼ year
2.	?	$1250	2% per yr.	6 months
3.	?	$1600	3% per yr.	2 yrs. 2 mos.
4.	?	$80	2% per mo.	½ year
5.	?	$4500	7½% per yr.	30 months

6) Mr. Jennings has $200 in a savings bank which pays 5% simple interest per year. How much interest does he receive each 1/2 year? What would be the amount after 5 years?

7) Mr. Jennings also has a loan of $200 in a commercial bank and he pays 1/2% (1/2%=.5%) simple interest per month. If he pays the loan back in <u>1 year</u>, how much interest does he pay? What amount?

8) If you have $700 to put in a savings bank for 2 years, which bank would you choose of the three below?

Simple Interest	= Principal	x Rate	x Time
Bank A ?	\$700	$\frac{1}{4}$% per mo.	24 months
Bank B ?	\$700	4% per yr.	2 years
Bank C ?	\$700	$1\frac{1}{2}$% per $\frac{1}{2}$yr.	2 years

OBJECTIVE 7-57
DEPOSIT SLIPS AND CHECKS

OBJECTIVE:

The student will be able to make out deposit slips and write checks.

PRETEST

A) Make out deposit slips in your name for each of the following deposits in the State Bank of Boston, Mass. Use today's date.

1) You deposit $54.25 in cash, a check drawn on the Sixth National Bank for $120.35, and a check drawn on Local Trust Company for $38.26. Your account number is 796-18237.

2) You deposit $28.27 in cash, a check drawn on Federal Savings Bank for $10.30, and a check drawn on Township Trust for $19.78. Use tomorrow's date and the same account number.

B) Write a check for each of the following drawn on your account at State Bank:

1) A week after your last deposit, you must pay Sears and Co. $61.26 for furniture you purchased. Fill in the stub.
2) The next day, you receive your gas and electric bill for $30.55. Use a check to pay this bill. Make it out to Massachusetts Electric Company. Fill in the stub. What is the balance carried forward?

A deposit slip is filled out by the depositor for each deposit. Each deposit slip shows the name and the account number of the depositor, the date, the items deposited and the total amount of the deposit.

EXAMPLE:

Make a deposit slip consisting of $150.25 cash, a check for $60.39 drawn on the Mutual Bank and a check for $70.38 drawn on the Liberty Bank.

STATE BANK
Boston, Mass.

Date: October 3, 1977

Please print or type
Name of account and number.

JOHN SMITH

Your Account Number: 876-2395

All checks deposited subject to final payment		
Cash:	$150	25
Checks:		
1. Mutual Bank	60	39
2. Liberty Bank	70	38
3.		
4.		
Total:	$281	02

EXERCISE:

Make out a deposit slip to a bank near your home. You deposit $236.58 in cash, a check from Fitchburg Savings Bank for $20.42, and a check for $132.25 from Savings and Loan Bank. What is the total? Use today's date, your name, and make up an account number of your own.

A check is used to pay for items purchased, or to pay your bills and taxes. Each check shows the check number, the date, the name of the payee, the amount and the drawer's signature (your name). A check stub is always filled out when a check is written. The stub gives the number of the check, the amount, the date, the name of a person or business to whom the check was written, the reason for the check, the total, and the balance carried forward.

EXAMPLE:

Write a check for Smith Hardware Store for $10.76 to pay for tools purchased.

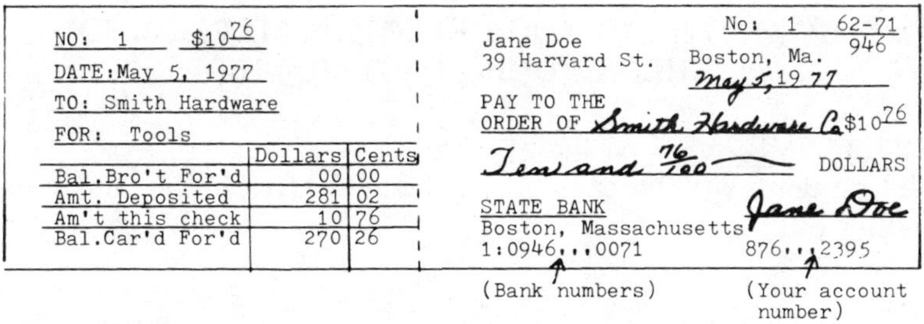

EXERCISE:

Write a check drawn on the bank you chose to deposit your money in for the exercise on page 114. Make the check payable to the New England Telephone Company in the amount of $15.75. Use today's date, the same account number as before and your amount deposited in the previous problem.

POSTTEST

1) On October 20, 1977, you deposit $84.22 in cash, a check drawn on the First Bank of Boston for $44.97, and a check drawn on the Workers Credit Union for $92.00. Write a deposit slip in your name and the above date to show this deposit in a bank near your home. Make up your own account number.
2) Make a check drawn on the bank you chose above to pay for a coat you bought from A-Mart that cost $45.92. Use today's date and the same account number used above and the amount deposited above as the amount deposited on the check stub. What is the balance carried forward?
3) Using tomorrow's date, write another check for $54.27 to pay your excise tax bill. Make it payable to the Commonwealth of Massachusetts. Use the same account number as above and use the balance carried forward from problem number 2 above as the balance brought forward on the stub for this check. Note that you have made no new deposits.
4) How much money do you have left in the bank? How much would you have if you deposit $250.00 more?

OBJECTIVES 7-(60,61)

HOW TO READ YOUR ELECTRIC METER AND CHECK YOUR ELECTRIC BILL

OBJECTIVE:
The student will be able to read an electric meter and check his/her bill.

PRETEST

1) SUBTRACT:

a) 54732 − 53932

b) 68636 − 67843

c) 46300 − 45940

d). 47899 − 46956

2) Find the readings of these meters:

3) On the first of June, Mr. Davis' electric meter reading was 64107. In July, the reading was 64743. If he pays a fuel cost of .024/kWh, how much does he have to pay?

The electric meter has five dials. Each dial is numbered 0 to 9 and has one hand. What you see is something like this:

Notice that every other dial is numbered backwards. Each dial is read as the number the hand has just passed, so

 is read as <u>5</u> and  is read as <u>3.</u>

This set of dials would be read as 52580:

The amount of electricity that is used is the <u>difference</u> of <u>two readings.</u> The difference of the two dials below is the amount of electricity used.

The amount of electricity used is 14483 −14205 or 278 kilowatt hours (kWh). Electricity is always measured in kilowatt hours (kWh).

HOW TO CHECK YOUR ELECTRIC BILL

On your electric bill, there are the two readings and the dates on which they were taken, the amount of electricity in kWh that has been used in that time period (usually one month). On the bill, there is a rate code which tells how much the electricity costs. The bill includes the amount due and may also have an electric fuel charge. An electric bill may look something like this:

Date From	Date To	Prev. Read.	Pres. Read.	kWh	Rate Code	Amount
10/07	11/09	41480	42236	756	10	$43.36
Electric Fuel Cost			.02132/kWh		$16.12	

To check the bill, first check the electric fuel cost. Multiply .02132 x 756 = 16.11792. Rounding this off we get $16. 12, which checks.

Subtract $16.12 from the amount of the bill:
$43.36 − $16.12 = $27.24.

The rate code is 10 and on the back of the bill is the description of the rate code:

First 15 kWh or less— $2.00
Next 35 kWh @ 5.7¢ per kWh
Next 50 kWh @ 4.7¢ per kWh
Next 100 kWh@ 3.7¢ per kWh
Next 550 kWh @ 3.1¢ per kWh
Over 750 kWh @ 2.4¢ per kWh

The first 15 kWh is \$2.00

The next 35 kWh is (35 x 5.7¢) or (35 x \$ .057)

35 x .057 = 1.995 or \$2.00 (rounding off)

The next 50 kWh is (50 x \$.047)

50 x .047 = \$2.35

The next 100 kWh is (100 x \$.037)

100 x .037 = \$3.70

The next 550 kWh is (550 x \$.031)

550 x \$.031 = \$17.05

Add \$2.00 + \$2.00 + 2.35 + \$3.70 + 17.05 = \$27.10

So, *the first 750 kWh cost \$27.10.*

Every Kwh over 750 kWh costs 2.4 ¢ or \$.024 per kWh,

so 756 (our usage in kWh) −750 = 6

6 x \$.024 = \$.144 or \$.14 (rounding off)

THE TOTAL COST FOR 756 kWh IS \$27.10 + \$.14 or \$27.24.

Since \$43.36 − \$16.12 = \$27.24, the bill is correct.

Using the same rating code and fuel cost in the example above, solve the following problems:

1) Check a bill for \$22.53 for 357 kWh.
2) What is the cost for 305 kWh?
3) How many kWh were used if the bill is \$21.63 and the fuel charge is \$7.25?
4) Is this bill correct: \$28.47 for 425 kWh?

POSTTEST

1) Find the readings of these meters:

a)

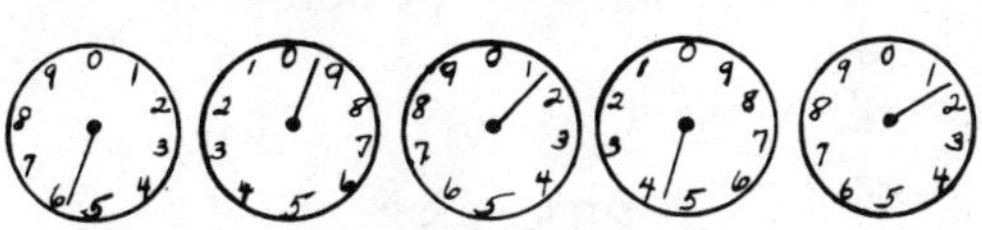

b)

2) If Mr. Millan's last month meter reading was 47854 and this month's reading is 48414 and his fuel cost and rate are the same as the ones used in the example, how much is his bill?
3) Here is a sample bill. See if it is correct.

Date From	Date To	Prev. Read.	Pres. Read.	kWh	Rate Code	Amount
1/07	2/08	74893	75755	862	12	$63.63
Electric Fuel Cost			.03114/kWh		$26.84	

Rate Code 12
First 20 kWh is $2.50
Next 30 kWh @ 6.2¢ per kWh
Next 50 kWh @ 5.4¢ per kWh
Next 100 kWh @ 4.7¢ per kWh
Next 550 kWh @ 3.9¢ per kWh
Next 750 kWh @ 3.2¢ per kWh

4) Find a bill and bring it in to check.

OBJECTIVES 7-(62,63)

HOW TO READ YOUR GAS METER AND CHECK YOUR GAS BILL

OBJECTIVES: The student will be able to:
Read a gas meter.
Check a gas bill.

PRETEST

SUBTRACT:

1) a) 2650 − 2613 b) 6743 − 6532 c) 4321 − 3927 d) 8473 − 8324 e) 2478 − 2222

2) Find the readings of the following meters:

a)

b)

3) If the first reading is 2613 and the second reading is 2650, how many cubic feet of gas have been used if the meter reads in hundreds only?

Every gas meter has two rows of dials. The bottom row of dials is for testing purposes only, so do not read these. Read only the top row of dials. The top row has either three or four dials. Your meter may look something like this:

Each dial is numbered from 0 to 9 and each has a little hand like a clock. Your meter may look something like this:

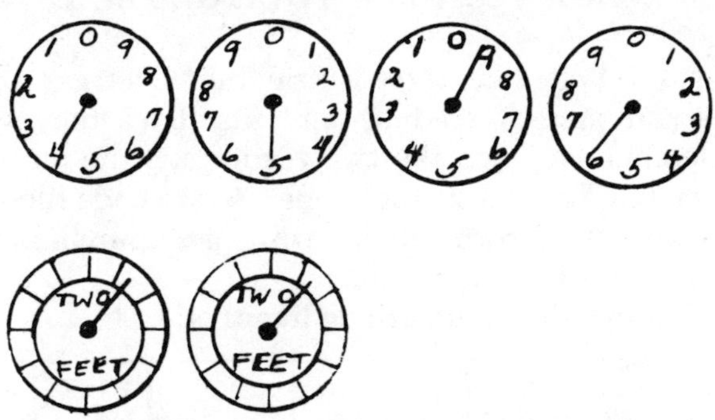

Every other dial goes in the opposite direction of the other. When reading the dials, read the last number passed by the hand.

This dial is read as *5*

This dial is read as *3*

The amount of gas that is used is the difference of the two readings taken at two different times.

613

650

Subtract the first amount (reading) from the second amount (reading) and you get 37. Since the meter reads (or is read) only in *hundreds*, the amount is *37 hundreds* or *37CCF (CCF means 100 cubic feet).*

HOW TO CHECK YOUR GAS BILL

The gas bill is like the electric bill. The gas bill has the previous and present readings and the dates they were taken. The difference between the two readings is the number of 100 cubic feet (CCF) used in that time. A rate code like that of the electric bill is shown along with gas purchasing cost to determine the total amount due.

A gas bill may look something like this:

Date From	Date To	Prev. Read.	Pres. Read.	100 Cu.Ft.	Rate Code	Amount
2/01	3/04	9512	9632	120	10	$27.95
Purchasing Gas Cost .0477/CCF $5.72						

To check the bill, first check the gas purchasing cost.
.0447 (gas purchase cost) x 120 (amount of gas in CCF) = 5.724.
Round off 5.724 = $5.72.
Subtract $5.72 from the amount of the bill:

$27.95 – $5.72 = (amount of bill – gas purchase cost) = $22.23.

The *rate code is 10.* The description of the rate code is on the back of the bill:

First 1 CCF — $2.00
Next 9 CCF @ 27¢ per CCF
Next 20 CCF @ 19¢ per CCF
Next 50 CCF @ 16¢ per CCF
Next 140 CCF @ 15¢ per CCF
Excess @ 14¢ per CCF

The amount of gas used is 120 CCF.
The first 1 CCF cost $2.00
The next 9 CCF cost ($.27 x 9) or $2.43
The next 20 CCF cost ($.19 x 20) or $3.80
The next 50 CCF cost ($.16 x 50) or $8.00
The next 40 CCF cost ($.15 x 40) or $ 6.00

The total of 120 CCF cost:

1	CCF	$2.00
9	CCF	2.43
20	CCF	3.80
50	CCF	8.00
40	CCF	6.00
120	CCF	$22.23

The bill is correct, because $27.95 (total cost) – $5.72 (purchase cost) = $22.23

POSTTEST

1) Give the readings of these meters:

a) 

b) 

2) Use this gas purchasing cost and this rate for the following problems:
First 1 CCF— $2.00
Next 9 CCF @ $.27 per CCF
Next 20 CCF @ $.19 per CCF
Next 50 CCF @ $.16 per CCF
Next 140 CCF @ $.15 per CCF
Excess @ $.14 per CCF

Gas purchase cost .0502/CCF

a) If you use 237 CCF of gas in one month, how much will you have to pay?

b) Check this sample bill. Is it correct?

Date From	Date To	Prev. Read.	Pres. Read.	CCF	Rate Code	Amount
1/07	2/08	1234	1474	240	10	$52.08
Gas purchase cost		.0502/CCF				

3) Find an old gas bill at home and check it to see if it is correct. Bring in the bill and the work to your teacher.

OBJECTIVE 7-93

USING EXPONENTS TO MULTIPLY

OBJECTIVE:

The student will be able to use exponents to perform simple multiplication ($2^3 \times 3^2$).

PRETEST

1. An *exponent* tells us how many times to use a number as a ______________.
2. The number being used as a factor is called the __________.

3. $2^3 \times 3^2 =$
4. $3^3 \times 2^2 =$
5. $(4^2)(4^2) =$

You have worked with exponents before, so you should know that an exponent is a *small* number placed to the right and above another number. The other number, which is a regular size number, is called the *base.*

EXAMPLE: 5^3

3 is the *exponent*

5 is the *base*

The 3 tells us that 5 is to be used as a factor *three* times. Another way to write this statement would be:

5 x 5 x 5 (*Answer*: 125)

NOW YOU TRY TO DO SOME EXAMPLES.

1. 4^3 = ______ x ______ x ______ = ______
2. 8^2 = ______ x ______ = ______
3. 2^4 = ______ x ______ x ______ x ______ = ______

YOU DO NOT HAVE TO FIND THE ANSWERS FOR THESE EXAMPLES. JUST WRITE OUT THE SECOND STEP:

4. 100^8
5. 14^3
6. 25^2
7. $2^4 \times 5^2$
8. $3^2 \times 3^3$
9. $15^2 \times 2^4$
10. 1×5^{13}

We can make the multiplication *easier*, if we are multiplying two or more number which have the *same base.*

EXAMPLES: $2^2 \times 2^3 \times 2^1 = 2^6$

$3^4 \times 3^{21} = 3^{25}$

How were the exponents *6* and 25 obtained? That's right, we *added* all of the exponents.

We could do this to make the work easier, because the *bases were the same* in both examples.

EXERCISE:

Write these examples using one exponent:

1) $2 \times 2^2 \times 2^3$ (Note: if the number has no exponent written, the exponent is always 1.)
2) $3^3 \times 3^2 \times 3^4 =$
3) $10^2 \times 10^2 \times 10^2 =$
4) $25^2 \times 25^2 \times 25^5 =$
5) $1 \times 1^2 \times 1 \times 1^{10} =$

POSTTEST

Describe these terms in your own words:

1) exponent 2) factor 3) base (when used with an exponent)
4) $5 \times 5^3 \times 5^4 \times 5 = 5^{\_\_\_}$.
5) $1^2 \times 1^6 \times 1 = 1^{\_\_\_} =$ ____.
6) What conclusion could you come to as a result of doing example 5?
7) *Find the answers:*

a) 2^5	b) 3^3	c) 4^5	d) 10^5
e) 1^{123}	f) 7^2	g) 2×3^3	h) $3^3 \times 3^2$
i) 4×6^2	j) $2^2 \times 3^2 \times 5$	k) $3^2 \times 8^3$	l) 25^3

OBJECTIVES 7 - (100-103)

METRIC LINEAR MEASURE

OBJECTIVES:

The student will be able to:
Define the prefixes used in the Metric System.
Express metric measures as powers of 10.
Use linear measure in the Metric System.
Solve problems involving linear measure in the Metric System.

It would be a good idea to review the packet covering Objectives 6-(86-88) before continuing. The packet covers some of the prefixes used in the Metric System and how to express them as powers of 10.

PRETEST

1. Give the prefixes that mean:
   a) one hundred b) one tenth c) one thousand
   d) one thousandth e) one hundredth f) ten
2. What is the basic unit of measure used in the metric system?
3. Use the powers of 10 to express the following:
   a) millimeter b) meter c) hectometer d) decimeter
   e) centimeter f) decameter g) kilometer

If you were able to do the pretest, you have a good start. If not, you will have to study this packet harder. You must know the following prefixes and how to express numbers as powers of 10. (You will find at least six prefixes that you have not studied before.)

PREFIXES USED TO EXPRESS *WHOLE NUMBERS*

Prefix	Abbreviation	Power of Ten	Multiple of a Meter
tera-	T	10^{12}	One *trillion* times a meter
giga-	G	10^9	*One billion* times a meter
mega-	M	10^6	*One million* times a meter
kilo-	km.	10^3	*One thousand* times a meter
hecto-	hm.	10^2	*One hundred* times a meter
deca-	dam. (or dkm.)	10^1	*Ten* times a meter

PREFIXES USED TO EXPRESS *PARTS OF A METER (FRACTIONS)*

Prefix	Abbreviation	Power of Ten	Multiple of a Meter
deci-	dm.	10^{-1}	*One tenth* of a meter
centi-	cm.	10^{-2}	*One hundredth* of a meter
milli-	mm.	10^{-3}	*One thousandth* of a meter
micro- micron	μ	10^{-6}	*One millionth* of a meter

nano-	10^{-9}	*One billionth* of a meter
pico-	10^{-12}	*One trillionth* of a meter

Can You Find The Words?

C	D	O	S	M	V	E	D	U	C	A	T	I	O	N
O	E	P	R	I	D	E	E	V	E	R	B	O	T	H
N	C	Q	P	N	V	I	C	A	R	M	O	Q	R	E
S	A	W	A	B	C	K	D	P	T	N	P	S	T	C
I	M	I	L	L	I	M	E	T	E	R	U	V	T	E
D	M	A	T	H	K	O	C	S	Y	M	B	O	L	N
E	Q	U	A	L	I	B	I	S	U	M	A	D	D	T
C	A	B	K	I	L	O	M	E	T	E	R	C	D	I
A	H	E	C	T	O	M	E	T	E	R	S	U	B	M
M	I	K	I	L	M	E	T	E	R	L	O	V	E	E
E	G	O	T	M	E	T	E	R	F	R	E	E	N	T
T	H	R	X	M	T	A	R	M	E	T	R	I	C	E
D	E	C	A	M	E	T	E	R	Q	L	O	C	H	R

Find the words in the box that define the terms below. Circle the words as you find them. Then, write the correct word by each term.

1.) 39.37 inches ________
2.) $\frac{1}{100}$ meter ________
3.) 1000 x a meter ________
4.) 10^{-1} meter ________
5.) 10 x a meter ________
6.) 10^{-3} meter ________
7.) 100 x a meter ________

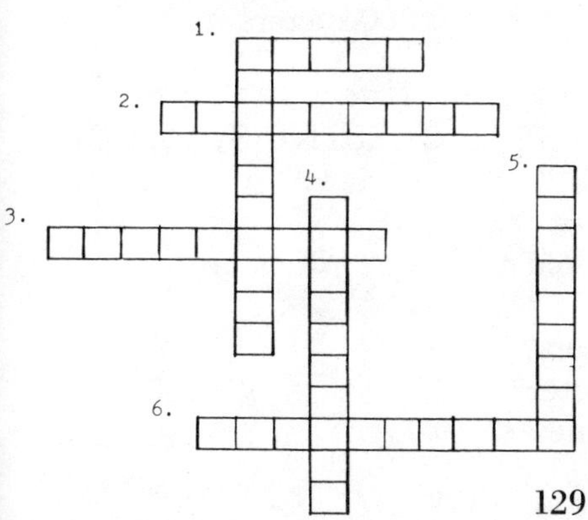

Try to do this without looking at your chart of terms. This will help you to learn the terms. Think of the prefixes.

ACROSS		DOWN	
1.	39.37 inches	1.	1/1000 meter
2.	1000 x a meter	4.	1/100 meter
3.	10 x a meter	5.	1/10 meter
6.	100 x a meter		

Now, why don't you try to make up one of your own? If you wish, you can make it more difficult by using positive and negative powers of 10. For example, if you want to use the word "micron," write 10^{-6}.

If you make up a good one, you can ask your friends to solve it.

MEASURES OF LENGTH

10 millimeters (mm.)	=	1 centimeter (cm.)
10 centimeters (cm.)	=	1 decimeter (dm.)
10 decimeters (dm.)	=	1 meter (m.)
10 meters (m.)	=	1 decameter (dam.) (also, de*k*ameter)
10 decameters (dam.)	=	1 hectometer (hm.)
10 hectometers (hm.)	=	1 kilometer (km.)

1 centimeter (cm.)	=	10 millimeters (mm.)
1 meter (m.)	=	10 decimeters (dm.)
	=	100 centimeters (cm.)
	=	1000 millimeters (mm.)
1 kilometer (km.)	=	1000 meters (m.)

NOW, TRY TO ANSWER THESE QUESTIONS WITHOUT LOOKING AT THE TABLE:

Which of the above express whole numbers? Which are fractions?
A decameter is how many meters?
How many centimeters in a meter?
A meter is what part of a kilometer?

A decameter is what part of a hectometer?
How many decimeters are there in a meter?
Which is larger: hectometer or millimeter?
decimeter or decameter?
centimeter or hectometer?
millimeter or kilometer?
How many meters are there in a hectometer?
How many centimeters are there in a decimeter?
How many millimeters are there in a decimeter?

If you know how to multiply and divide by powers of 10, you will be able to work very easily with the Metric System. All you will need to do will be to move the decimal point the correct number of places to the right or the left. Learn how to do this by studying the following pages carefully.

CHANGING TO SMALLER UNITS OF MEASURE

To change from a *larger* unit of measure to a *smaller* unit of measure, *multiply*. (You will have more of the smaller units.) For example, to change:

Dollars to cents, multiply the number of dollars by *100*.
Years to days, multiply the number of years by 365.
Miles to feet, multiply the number of miles by 5,280.
Meters to decimeters, multiply the number of meters by 10.
Meters to centimeters, multiply the number of meters by 100.
Meters to millimeters, multiply the number of meters by 1000.
Meters to microns, multiply the number of meters by 1,000,000.

REMEMBER that to *multiply* by powers of 10, all you have to do is move the decimal point *one place to the right for each zero.*

To multiply by 10, move the decimal point *one* place to the right.

To multiply by 100, move the decimal point *two* places to the right.

To multiply by 1000, move the decimal point *three* places to the right.

To multiply by 1,000,000, move the decimal point *six* places to the right.

EXERCISE:

Multiply the following by moving the decimal point the necessary number of places to the right:

1) 1000 x 134__________
2) 10 x 435.9__________
3) 10^2 x 23.45__________
4) 10^6 x 67__________
5) 100 x 432.1__________
6) .086754 x 10^2 __________
7) 439.7654 x 10^3 __________

Before you go any further, you should be able to multiply using powers of 10. Remember that, if the number is expressed by a 10 with an exponent, the exponent tells you how many zeros there are after the 1. Thus, 10^4 tells you there are 4 zeros, so the number is 10,000.

Now, why don't you make up some examples of your own and do them? If you do not want to work alone, you could try them out on a friend.

CHANGING TO LARGER UNITS OF MEASURE

To change from a *smaller* unit of measure to a *larger* unit of measure, *divide*. (It takes several of the smaller units to equal one of the larger units.) For example, to change:

Cents to dollars, divide the number of cents by 100.
Days to years, divide the number of days by 365.
Feet to miles, divide the number of feet by 5,280.
Decimeters to meters, divide the number of decimeters by 10.

Centimeters to meters, divide the number of centimeters by 100.
Millimeters to meters, divide the number of millimeters by 1000.
Microns to meters, divide the number of microns by 1,000,000.

REMEMBER that to divide by powers of 10, all you have to do is move the decimal point *one place to the left for each zero.*

To divide by 10, move the decimal point *one* place to the left.
To divide by 100, move the decimal point *two* places to the left.
To divide by 1000, move the decimal point *three* places to the left.
To divide by 1,000,000, move the decimal point *six* places to the left.

Divide the following by moving the decimal point the necessary number of places to the left. When multiplying by negative powers of 10 (10^{-2}), go left as in division. This is because when you multiply by a fraction with a numerator of 1, you will get the same answer by dividing by the denominator. For example, $165 \times 10^{-2} = 165 \times 1/10^2 = 165 \times 1/100 = 165/100 = 1.65$

EXERCISE:

1. 7500 ÷ 10_______
2. 23.456 ÷ 100_______
3. 465,029 ÷ 1000_______
4. 32 ÷1000_______
5. 326,896 ÷ 1,000,000_______

6) 36×10^{-1}_______
7) 1436.2×10^{-4}_______
8) .0231 ÷ 100_______
9) $.29 \times 10^{-6}$ _______
10) 111.11111×10^{-3}_______

Now, make up a few examples of your own. Make sure you can do this before continuing.

Do the following exercises by moving the decimal point to

multiply or divide. First, think whether you are changing to *larger or smaller* units of measure. Then, decide whether to *multiply or divide.*

EXAMPLES:

Change 800 meters to hectometers.
100 meters = 1 hectometer
Hectometers are larger units than meters, so *divide by 100.*
Move the decimal point *two places to the left.*
Answer 800 meters = 8 hectometers.

Change 4 kilometers to centimeters
1 meter = 100 centimeters and
1 kilometer = 1000 meters, so
1 kilometer = 1000 x 100 = 100,000 centimeters
Centimeters are smaller units than kilometers, so *multiply* by 100,000. Move the decimal point *5 places to the right.*
Answer 4 kilometers = 400,000 centimeters.

Change 4 kilometers 8 meters to meters
1 kilometer = 1000 meters, so
8 meters = .008 kilometers and
4 kilometers 8 meters = *4.008* kilometers
Meters are smaller units than kilometers, so *multiply* by 1000. Move the decimal point *three places to the right.*
Answer: 4 kilometers 8 meters = 4008 meters.

EXERCISE:

1. Change to hectometers:	450 m.	2498 m.	65 m.
2. Change to kilometers:	16.75 hm.	736 hm.	19 hm
3. Change to millimeters:	5 km.	.5 km.	16 km
4. Change to kilometers:	6,000,000 mm.	25,000 mm.	
	250,000 mm.		

5. Fill in the blanks:
   1 m. 8 cm. =____________m.
   3 cm. 2mm. =____________mm.
   5m. 8 dm. 5 cm. = ____________cm.
   7 cm. 3 mm. = ____________cm.
   4 m. 7cm. 2 mm. =____________m.

POSTTEST

1. Write any 5 prefixes and give the power of 10 used to express each one.
2. Fill in the blanks:
   a) 7m. 61 cm. = ____________m.
   b) 15 cm. 8 mm. = ____________cm.
3. If an athlete jumps 1 meter 95 centimeters, what is the height of his jump in meters?
4. If a circle has a diameter of 34 meters, what is its area? (Remember: $A = \pi r^2$. What part of the diameter is the radius?)
5. What is the area of a square 5 mm. on a side? ($A = s^2$)
6. Find the perimeter of a rectangle 16 meters long and 13 meters wide.
7. The winner at a track meet threw the discus 55.82 meters. If the runner-up throw was 54 meters 27 centimeters, what was the difference between the two throws?
8. How many kilometers can an airplane fly in one minute, if the speed of the plane is 135 meters per second?
9. How many yards of material are there in 2 meters? (1 meter = 39.37 inches. Carry answer out two places.)
10. Find your height in meters and decimeters. Ask a friend to help you.

OBJECTIVES FOR LEVEL 8

The student will be able to:

8-1 Add whole numbers, renaming addends mentally to get convenient combinations such as multiples of 10.

8-2 Subtract whole numbers involving the regrouping of digits in the minuend.

8-3 Multiply whole numbers by a single-digit multiplier without writing partial products.

8-4 Divide whole numbers up to 3-digit divisors.

8-5 Apply the Commutative Properties of Addition and Multiplication to whole numbers.

8-6 Apply the Associative Properties of Addition and Multiplication to whole numbers.

8-7 Apply the Distributive property of Multiplication to whole numbers.

8-8 Apply the rules for the Order of Operations to real numbers (parentheses, brackets, multiplication and division, addition and subtraction).

8-9 Apply the rules for Divisibility.

8-10 Insert the correct symbols in a number sentence to make it a true statement.

8-11 Insert the correct symbols in an inequality to make it a true statement.

8-12 Identify various types of rational numbers (proper fractions, improper fractions, unit fractions, equivalent fractions, mixed numbers).

8-13 Read rational numbers (in the form of a fraction) correctly.

8-14 Use fractions to express ratios.

8-15 Use the symbols for "greater than," "less than," and "equal to" when comparing fractions.

8-16 Use cross-multiplication to compare fractions.

8-17 Determine whether a statement comparing two ratios is a proportion or an inequality.

8-18 Solve problems involving ratio and proportion.

8-19 Write one or more equivalent fractions for a given fraction.

8-20 Identify the L.C.M. (Least Common Multiple) of two or more numbers.

8-21 Add fractions with unlike denominators.
8-22 Subtract fractions with unlike denominators.
8-23 Identify the G.C.F. (Greatest Common Factor) of two or more numbers.
8-24 Multiply two or more fractions.
8-25 Multiply whole numbers by fractions.
8-26 Multiply two or more mixed numbers.
8-27 Divide whole numbers by fractions and vice versa.
8-28 Divide fractions by fractions.
8-29 Divide fractions by mixed numbers and vice versa.
8-30 Apply the Commutative Properties of Addition and Multiplication to fractions.
8-31 Apply the Associative Properties of Addition and Multiplication to fractions.
8-32 Apply the Distributive Property of Multiplication to fractions.
8-33 Solve problems involving the addition and subtraction of fractions.
8-34 Solve problems involving the multiplication and division of fractions.
8-35 Find the fraction half-way between two other fractions.
8-36 Name the place value of each digit in a decimal numeral up to the millionths place.
8-37 Use exponents to express decimal numerals in expanded form.
8-38 Read numbers containing decimal numerals up to the millionths place.
8-39 Add and subtract decimal numerals up to the millionths place.
8-40 Multiply decimal numerals up to the millionths place.
8-41 Divide decimal numerals, using divisors up to the ten-thousandths place.
8-42 Change decimal numerals to fractions—(terminating).
8-43 Change fractions to decimal numerals.
8-44 Solve problems involving decimal numerals.
8-45 Change decimal numerals to percents.
8-46 Change percents to decimal numerals.

8-47 Express percents as fractions.
8-48 Express fractions as percents.
8-49 Use the terms *base, rate, and percentage.*
8-50 Find a percent of a given number.
8-51 Find what percent one number is of another.
8-52 Find the whole (100%) when a part of it is given.
8-53 Solve problems involving the three types of percent.
8-54 Use a table to find square roots to a given decimal place.
8-55 Find square roots by approximation. (Short method.)
8-56 Find square roots by computation. (Long method.)
8-57 Use the Pythagorean Theorem.
8-58 Distinguish between *rational* and *irrational* numbers.
8-59 Recognize *terminating, repeating, and non-terminating non-repeating* decimal numerals.
8-60 Express common fractions as *repeating* or *terminating* decimal numerals.
8-61 Express decimal numerals as common fractions.
8-62 Apply the Properties of Zero to the Real Number System.
8-63 Apply the Properties of One to the Real Number System.
8-64 Use integral powers of 10 to find products
8-65 Use integral powers of 10 to find quotients.
8-66 Use large and small numbers in scientific notation.
8-67 Identify: Point, line, half-line, line-segment, and ray.
8-68 Identify: Complementary, supplementary, adjacent and vertical angles.
8-69 Identify: Interior, alternate-interior, exterior and corresponding angles.
8-70 Identify: Parallel and perpendicular lines.
8-71 Name polygons by the number of their sides.
8-72 Determine whether geometric figures are congruent.
8-73 Recognize and illustrate line and point symmetry.
8-74 Find the perimeters of various polygons.
8-75 Find the circumference of a circle.
8-76 Find the area of a square, rectangle and parallelogram.
8-77 Find the area of a triangle.
8-78 Find the area of a trapezoid.
8-79 Derive formulas for finding the volumes of various prisms and pyramids from the formula V = Bh.

8-80 Find the volumes of rectangular prisms and of various types of pyramids.
8-81 Find the volumes of cylinders and cones.
8-82 Find the volume of a sphere.
8-83 Find the surface areas of various polyhedra.
8-84 Identify parts of a circle: Radius, diameter, circumference, arc, chord, central angle.
8-85 Find the greatest possible error for measurement.
8-86 Find the greatest possible round-off error for measurement.
8-87 Add and subtract integers.
8-88 Multiply and divide integers.
8-89 Translate worded mathematical sentences into the language of Algebra and vice-versa.
8-90 Recognize the distinction between algebraic expressions and equations.
8-91 Solve simple algebraic equations involving the sum or difference of two terms.
8-92 Solve simple algebraic equations involving the product or quotient of two terms through the use of inverse operations.
8-93 Find the solution set of an equation involving two variables, given the replacement set.
8-94 Find the solution set of an inequality involving two variables, given the replacement set.
8-95 Determine the probability of an event.
8-96 Construct bar, circle, and line graphs.
8-97 Construct frequency tables and histograms.
8-98 Find the following measures of central tendency: Range, mean, median, mode.
8-99 Distinguish between the X axis and the Y axis.
8-100 Write the signs for each quadrant of a graph.
8-101 Determine the number of subsets in a given set (2^n).
8-102 List all of the subsets of a given set.
8-103 Determine the union, intersection and disjunction of sets.
8-104 Graph a given set of ordered pairs.

8-105 Write a given numeral in any base in expanded notation using exponents.
8-106 Write as a numeral an expanded numeral in any given base.
8-107 Find the value of any digit in a given numeral in any given base.
8-108 Change a numeral from Base 10 to any given base.
8-109 Change a numeral from any given base to Base 10.
8-110 Perform addition and subtraction in a given base.
8-111 Perform multiplication and division in a given base.
8-112 Apply correctly the prefixes used in the Metric System (*deci-*, *centi-*, *milli-*, *kilo-*, *deka-*, *hecto-*,)
8-113 Define the various units of linear measure in the Metric System.
8-114 Define the various units of weight in the Metric System.
8-115 Define the various units of capacity in the Metric System.
8-116 Change to larger or smaller units of measurement accurately by applying the rules for multiplying and dividing by powers of 10.
8-117 Apply knowledge of the Metric System to other disciplines (Home Arts, Industrial Arts, Science, etc.)
8-118 Round off decimal numerals to any given place value.

OBJECTIVE 8-8

ORDER OF OPERATIONS

OBJECTIVE:

The student will be able to apply the rules for the Order of Operations to real numbers (parentheses, brackets, multiplication and division, addition and subtraction.)

PRETEST

Simplify each of the following expressions:

1) 11 + 2 (6 + 4) − 3(1 + 3)
2) 27 + 10 − (11 + 3)
3) 12 x 6 ÷ 3 x 2 ÷ 48
4) (2 x 3 − 12 ÷ 3) ÷ 2
5) (13 x 5 − 3 x 5) ÷ (8 + 2)
6) − 24 ÷ 4 ÷ (−3)
7) [36 − 4 (3 + 1)] ÷ 4 − 6
8) [64 − 5(1 + 4)] ÷ (−8 −5)
9) 118 ÷ [9(7 −1) + 5]
10) 24 ÷ 6(−3) x 8 ÷ (−2)

If given this expression, 4 x 3 − 2 x 4, can you decide what operations should be done first? Should you multiply and then subtract or should you subtract and then multiply? You can see how this expression could become confusing, because there are different answers depending on which operations you choose to do first.

To avoid questions about what to do first in simplifying an expression, mathematicians have agreed on the following steps or rules:

1) First, simplify the expression within each *symbol of inclusion* (parentheses ( ), brackets [ ]).

2) Perform the operations of multiplication and division *in order from left to right.*
3) Finally, do the operations of addition and subtraction *in order from left to right.*

LET'S TRY A FEW EXAMPLES.

EXAMPLE 1: $(8 + 4 + 3) \div 3 - 2$

Step 1: Do the work in the parentheses: $15 \div 3 - 2$

Step 2: Do the division: $15 \div 3 = 5$

Step 3: Subtract: $5 - 2 = 3$

The answer is 3.

EXAMPLE 2: $18 \div 2 \times 3 - 5 - 20$

Since there are no *symbols of inclusion*, we go to the next step which is the multiplication and division *in order from left to right.*

Step 2: $18 \div 2 = 9$; $9 \times 3 = 27$

Step 3: $27 - 5 - 20 = 22 - 20 = 2$ (Addition and subtraction *from left to right)*

The answer is 2.

EXAMPLE 3:

$(24 - 3 + 9 \div 3) \div (8 \times 1/2)$

$(24 - 3 + 3) \div 4$

$(21 + 3) \div 4$

$24 \div 4$

$24 \div 4 = 6$

The answer is 6.

(Note: When doing work in *symbols of inclusion*, follow the rules for the *Order of Operations.)*

EXERCISES:

Simplify: (Note: Remember that no sign between numbers means multiply.)

1) $49 + 5 \times 0$
2) $14 \div 7 \times 2$
3) $80 \div 8 - 80 \div 10$
4) $(27 \div 9) + [(27 + 9) \div 3]$
5) $[(16 + 4) \div (3 \times 2)] \div 13$
6) $8 + (-2) + 4(-2)$
7) $(-7) \times 4 - 3 + 2 + (-4)$
8) $(-6)4 - 4(3 + 6)(5)$
9) $(-243) \div 3 - 4(-3)$
10) $[(-4(1) + 6(-4) - 4(6)] \div [(-4 \times 4) \div 16]$

EXTRA FOR EXPERTS:

When simplifying expressions involving variables, we use the same rules for the Order of Operations. Here is an example:

$4(-x + y) - 4x - 3y + y(7 + x)$ (Watch signs !!!)
$-4x + 4y - 4x - 3y + 7y + xy$ (*NOTE:* When working with *variables*, we can only combine *like terms.*)
$-8x + 8y + xy$ (Answer)

TRY THESE EXAMPLES:

1) $x + x(x + x) - x(x + x)$
2) $2a + 3a(a + 4a) - 5a$
3) $6z + 3z(z)$
4) $a + 3b(4c) + 6d$
5) $r + s - 6s(4t)$
6) $(xy - zy) \div (4y - 3y)$
7) $d \div -d - a + 2b(4d)$
8) $x(x) \div (x)(x) \div (x)$
9) $x^2 \div x \div x$
10) $x^6 \div x^3 \div x^2 \div x$

POSTTEST

SIMPLIFY THE FOLLOWING EXPRESSIONS:

1) $(-3) - 4(-3 + 2) + 4$
2) $8 - 9 \times 0$
3) $6 \times 1 + 9$
4) $4 \times 9 - 4 \times 1$
5) $3(11 + 5) \div 4 + 6$
6) $18 \div 2 \times 3 - 5 - 20$
7) $(5 \times 3 + 5 \times 4) \div (9 - 2)$
8) $(7 \times 7 - 3 \times 3) \div (7 - 3)$
9) $[6(2 + 4) - 1] \div (2 \times 3 + 1)$
10) $[29 - 5(3 - 2)] \div (2 - 1 + 7)$
11) $\left[\dfrac{12 + 8}{9 \times 6} - \dfrac{1}{54}\right] - \dfrac{1}{54}$
12) $\left[\left(\dfrac{50 + 25}{15}\right) \div 5\right] + 1$

13) (15 − 3 x 8 ÷ 2) ÷ 8 x 5
14) [(4 x 2 + 6 x 10 − 3 x 20) + 56] ÷ 13
15) 3 + 48 − 16 − 35 ÷ 7
16) ½ (8 ÷ 2 x 4) ÷ (8 x 4 ÷ 2)
17) 3.3 + 3(6) (5 − 1) − 2
18) (2 x 3 ÷ 3 − 1) ÷ (2 x 3 ÷ 6)
19) (4 x 5 x 20 − 6 ÷ 2) ÷ (13 + 3 x 1 x 5)
20) $\frac{1 + 44 \div 4 + 12 \times 44}{3 \times 3 - 3 \div 3 + 2}$

OBJECTIVE 8-19

EQUIVALENT FRACTIONS

OBJECTIVE:
The student will be able to write one or more equivalent fractions for a given fraction.

PRETEST

1) What is the key number in writing equivalent fractions?
2) What form must this number take?
3) Give two equivalent fractions for each of the following:

a)	1/2	=	___, ___	g)	1/4	=	___, ___
b)	3/5	=	___, ___	h)	5/9	=	___, ___
c)	3/4	=	___, ___	i)	2/3	=	___, ___
d)	4/9	=	___, ___	j)	2/7	=	___, ___
e)	4/15	=	___, ___	k)	8/11	=	___, ___
f)	1/3	=	___, ___	l)	5/12	=	___, ___

4) Rewrite the following fractions in lowest terms:

a) 5/10 b) 4/6 c) 7/21 d) 5/15 e) 9/27

f) 20/25 g) 8/56 h) 2/12 i) 3/36 j) 96/108

Equivalent fractions are fractions that have *different forms but the same value.*

To change a fraction to an equivalent fraction in higher terms (one having larger numbers in the numerator and denominator), *multiply by 1 in the form of a fraction:*

$$\frac{2}{2}, \frac{3}{3}, \frac{4}{4}, \frac{5}{5}, \frac{10}{10}, \text{etc.}$$

EXAMPLE:

Change $\frac{2}{3}$ to equivalent fractions having *higher terms.*

Step 1: Change 1 to a fraction: $\frac{5}{5}, \frac{2}{2}, \frac{12}{12}$, etc.

Step 2: Multiply one of these fractions by the given fraction:

$$\frac{2}{3} \times \frac{5}{5} = \frac{2 \times 5}{3 \times 5} = \frac{10}{15}$$

So, $\frac{2}{3}$ and $\frac{10}{15}$ are *equivalent fractions.*

$$\frac{2}{3} \times \frac{2}{2} = \frac{2 \times 2}{3 \times 2} = \frac{4}{6}$$

$$\frac{2}{3} \times \frac{12}{12} = \frac{2 \times 12}{3 \times 12} = \frac{24}{36}$$

So, $\frac{4}{6}$ and $\frac{24}{36}$ are *equivalent* to $\frac{2}{3}$

Remember the special property of 1: If you multiply any number by 1, you get the original number. Therefore, we can multiply any number by some form of 1, such as $\frac{5}{5}$, $\frac{7}{7}$, $\frac{9}{9}$, $\frac{3}{3}$, etc. and get for an answer a number that will have the same value as the original number. Remember, also, that when you are working with fractions, you must *multiply both the numerator and the denominator* of the fraction by *the same number*, in order not to change its value.

<u>EXERCISE:</u>

Find equivalents for the following fractions: (higher terms)

a)	1/2 x ____ =	g)	1/4 x ____ =
b)	2/3 x ____ =	h)	1/10 x ____ =
c)	5/8 x ____ =	i)	3/16 x ____ =
d)	1/3 x ____ =	j)	2/7 x ____ =
e)	3/4 x ____ =	k)	1/6 x ____ =
f)	5/6 x ____ =	l)	3/5 x ____ =

You can also find *equivalents* by changing fractions to *lower terms.* To change to *lower terms,* divide by a form of 1 as before. Be sure to *divide both the numerator and the denominator by the same number.*

EXAMPLE:

Reduce 4/6 to *lowest* terms. To get the fraction in *lowest terms, divide both the numerator and the denominator* by the *largest common factor* of each.

Step 1: The largest common factor of 4 and 6 is 2, so divide by 1 in the form of $\frac{2}{2}$

Step 2: $4/6 \div 2/2 = \frac{4}{6} \div \frac{2}{2} = \frac{2}{3}$

So, 4/6 reduced to lowest terms is $\frac{2}{3}$

Likewise, $\frac{10}{12} \div \frac{2}{2} = \frac{5}{6}$

$\frac{6}{9} \div \frac{3}{3} = \frac{2}{3}$

$\frac{12}{16} \div \frac{4}{4} = \frac{3}{4}$

EXERCISE:

Find three *equivalents* for each of the following fractions:

a) $\frac{12}{24} \div$ ________ =

b) $\frac{9}{18} \div$ ________ =

c) $\frac{20}{25} \div$ ________ =

d) $\frac{7}{49} \div$ ________ =

e) $\frac{10}{16} \div$ ________ =

f) $\frac{16}{31} \div$ ________ =

g) $\frac{4}{8} \div$ ________ =

h) $\frac{5}{25} \div$ ________ =

i) $\frac{9}{12} \div$ ________ =

j) $\frac{8}{56} \div$ ________ =

k) $\frac{9}{81} \div$ ________ =

l) $\frac{10}{40} \div$ ________ =

If the equivalents you found for the above fractions were not in lowest terms, change them so that they will be. For example, you could have had $\frac{9}{18} = \frac{3}{6}$. Change $\frac{3}{6}$ to $\frac{1}{2}$.

POSTTEST

1) Give two equivalents for each of the following fractions:

$\frac{1}{4}$, $\frac{1}{2}$, $\frac{1}{3}$, $\frac{2}{3}$, $\frac{3}{4}$, $\frac{1}{5}$, $\frac{3}{5}$, $\frac{1}{6}$, $\frac{5}{6}$, $\frac{3}{8}$, $\frac{5}{9}$, $\frac{7}{10}$

2) Use the denominators given to find equivalent fractions for the following:

a) $\frac{1}{2} = \frac{\quad}{8}$

b) $\frac{1}{3} = \frac{\quad}{15}$

c) $\frac{5}{7} = \frac{\quad}{21}$

d) $\frac{5}{9} = \frac{\quad}{72}$

e) $\frac{3}{4} = \frac{\quad}{24}$

f) $\frac{3}{16} = \frac{\quad}{48}$

g) $\frac{17}{32} = \frac{\quad}{64}$

h) $\frac{3}{8} = \frac{\quad}{40}$

i) $\frac{5}{12} = \frac{\quad}{60}$

j) $\frac{7}{8} = \frac{\quad}{32}$

k) $\frac{7}{9} = \frac{\quad}{45}$

l) $\frac{13}{16} = \frac{\quad}{32}$

m) $\frac{9}{32} = \frac{\quad}{96}$

n) $\frac{17}{24} = \frac{\quad}{48}$

o) $\frac{7}{10} = \frac{\quad}{100}$

p) $\frac{3}{4} = \frac{\quad}{100}$

q) $\frac{9}{20} = \frac{\quad}{100}$

r) $\frac{43}{50} = \frac{\quad}{100}$

s) $\frac{2}{25} = \frac{\quad}{100}$

t) $\frac{2}{5} = \frac{\quad}{10}$

u) $\frac{3}{8} = \frac{\quad}{100}$

OBJECTIVE 8-23

COMMON FACTORS

OBJECTIVE:

The student will be able to identify the G.C.F. (Greatest Common Factor) of two or more numbers.

PRETEST

USE COMMON FACTORS TO FIND THE PRODUCTS IN THE FOLLOWING EXAMPLES:

1) 1/5 x 1 1/9
2) 3/4 x 2/9
3) 2/3 x 5/6
4) 5 1/3 x 3/4
5) 9/10 x 4/27
6) 29/17 x 51/16
7) 35/12 x 9/14
8) 2/3 x 3/4 x 4/5
9) 5/8 x 34/15 x 9

DEFINITION:

The *Greatest Common Factor* is the *largest* number that can be *evenly divided* into two or more numbers.

EXAMPLE:

What is the Greatest Common Factor of 6 and 9?
3 is the largest number that can be evenly divided into both 6 and 9.

So, 3 is the Greatest Common Factor.
Sometimes, one of the numbers is the G.C.F.

EXAMPLE:

What is the G.C.F of 4 and 36?
4 is a factor of 36, so 4 can be evenly divided into itself and 36.

HOW TO USE COMMON FACTORS

Common factors are very helpful in the multiplication and division of fractions, because by using them you do not have to multiply large numbers. Also, your answer will be in lowest terms.

EXAMPLE: Multiply 4 9/10 x 2/7.

Step 1: Change 4 9/10 to an improper fraction: $\frac{49}{10}$

Step 2: Factor each term: 49 = 7 x 7; 10 = 5 x 2

Step 3: Write the factors as a fraction: $\frac{7 \times 7}{5 \times 2} \times \frac{2}{7}$

Step 4: Write both fractions as one: $\frac{7 \times 7 \times 2}{5 \times 2 \times 7}$

Step 5: Cross out common factors in the numerator and denominator:

$\frac{\cancel{7} \times 7 \times \cancel{2}}{5 \times \cancel{2} \times \cancel{7}}$ (Note: you can do this because $\frac{2}{2} = 1$ and $\frac{7}{7} = 1$)

So, 4 9/10 x 2/7 = 7/5 or 1 2/5.

EXAMPLE: 1 1/2 x 1 7/9

$$\frac{3}{2} \times \frac{16}{9} = \frac{3}{2} \times \frac{2 \times 8}{3 \times 3} = \frac{\cancel{3} \times \cancel{2} \times 8}{\cancel{2} \times \cancel{3} \times 3} = \frac{8}{3}$$

$\frac{8}{3} = 2\ 2/3$ (Answer)

EXERCISE: Use common factors to perform the following multiplications:

1) $\frac{1}{3} \times \frac{6}{7}$

2) $\frac{3}{4} \times \frac{16}{15}$

3) $\frac{1}{2} \times 6$

4) $\frac{1}{3} \times \frac{21}{22}$

5) $\frac{2}{7} \times \frac{1}{2} \times \frac{7}{8}$

6) $\frac{5}{12} \times \frac{12}{25} \times 1\ 4/5$

POSTTEST

USE COMMON FACTORS TO MULTIPLY EACH OF THE FOLLOWING:

1) 3/ 7 x 8/3
2) 7/9 x 15/14
3) 14/11 x 22/16
4) 24/17 x 51/36
5) 9/25 x 125/27
6) 16/49 x 343/64
7) 57/20 x 35/38 x 22/27
8) 67/8 x 5 1/3
9) How much will a trip cost, if you go 12 miles at 7 3/4¢ per mile?
10) A plane goes 600 miles per hour. At that rate, how far does it go in 6 1/4 hours?
11) Find the cost of 1½ dozen eggs at 88¢ per dozen.

OBJECTIVE 8-37

EXPANDED FORM

OBJECTIVE:

The student will be able to use exponents to express decimal numerals in expanded form.

PRETEST

1) Use exponents to simplify the following:
   a) 10 x 10 x 10
   b) 10 x 10 x 10 x 10 x 10
2) Write the decimal numeral for the following:
   a) 10^2
   b) 10^4
   c) 10^0
3) Write each of the following in expanded form in two ways:
   a) 232
   b) 34,758
   c) 93,210,435
4) Write the numerals expressed in expanded form:
   a) $(3 \times 10^4) + (7 \times 10^3) + (5 \times 10^2) + (2 \times 10^1) + (5 \times 10^0)$
   b) $(9 \times 10^6) + (3 \times 10^5) + (8 \times 10^3) + (4 \times 10^2) + (6 \times 10^1) + (2 \times 10^0)$
   c) $(1 \times 10^5) + (3 \times 10^4) + (6 \times 10^2) + (5 \times 10^1)$

To write a number in *expanded form*, *multiply each digit* in the number by its *place value*.

EXAMPLE:

Write 4795 in *expanded form*.
(Note: *Expand* means "stretch out" or "write longer." When we expand a number, we write it out in its longest form. 4795 is written in its shortest form.)
The place value of 5 is 1, so we write: (5 x 1)
The place value of 9 is 10, so we write: (9 x 10)
Likewise: 7 = (7 x 100) and 4 = (4 x 1000)
Notice that if we add up the values of all the digits, we have another way of expressing 4795:

$$(4 \times 1000) + (7 \times 100) + (9 \times 10) + (5 \times 1) =$$
$$4000 + 700 + 90 + 5 = 4795$$

This way of expressing a number is called expanded form or *expanded notation.*

EXAMPLE: Express 304,296 in two ways.

304,296 = (3 x 100,000) + (4 x 1000) + (2 x 100) + (9 x 10) + (6 x 1)

304,296 = 300,000 + 4000 + 200 + 90 + 6

Note: Only the *first* is *expanded form.* Notice, also that there is no term for the 10,000, because the digit in that place is zero.

EXERCISE: Write these numerals in expanded form:

1) 2583
2) 107, 693
3) 5,207,936
4) 74,102
5) 82,593,426

Did you find that you had to write quite a few zeros when you wrote the numerals in *expanded form?* There is a shorter way of writing numerals in *expanded form*, but to do this you must review *exponents.*

Remember that numbers multiplied together are called factors. *When the factors are the same,* we can use *exponents* to simplify the writing.

EXAMPLE:

Write 10 x 10, using an *exponent.*

10 is used as a factor *twice.* Use 2 for an exponent.

The number being used as a factor is called the *base.*

2 is the *exponent;* 10 is the *base.*

A short way to write 10 x 10 is 10^2.

10^2 is read "ten to the second power" or "ten squared."

10^3 is read "ten to the third power" or "ten cubed."

Other exponents are read:

10^4 = "ten to the fourth power"

10^0 = "ten to the zero power"

10^1 = "ten to the first power," etc.

NOTE:
Do not forget that any number raised to the zero power = 1. Remember also, that any number raised to the first power = the number itself.

EXAMPLES:
$12^0 = 1$; $12^1 = 12$; $85^0 = 1$; $1234^1 = 1234$, etc.

STUDY THE FOLLOWING TABLE OF THE POWERS OF 10:

10^0	=	1	
10^1	=	10	
10^2	=	100	(10 x 10)
10^3	=	1000	(10 x 10 x 10)
10^4	=	10,000	(10 x 10 x 10 x 10)
10^5	=	100,000	(10 x 10 x 10 x 10 x 10)
10^6	=	1,000,000	(10 x 10 x 10 x 10 x 10 x 10)

IF YOU STUDIED THE TABLE CAREFULLY, YOU WILL HAVE FOUND A PATTERN FOR THE POWERS OF 10. WHAT IS THIS PATTERN?

EXERCISE:

1) Fill in the blanks:
   a) To write 10^2, we write a 1 followed by ______ zeros.
   b) To write 10^5, we write a 1 followed by ______ zeros.
   c) To write 10^{100} we write a 1 followed by______ zeros.
   d) To write 10^0 we write a 1 followed by ______ zeros.
2) How many times is 10 used as a *factor?*
   a) 10^3 b) 10^4 c) 10^6 d) 10^{10}
3) Rewrite, using exponents:
   a) 10 x 10 x 10 x 10 x 10
   b) 10 x 10 x 10 x 10 x 10 x 10 x 10 x 10
4) Write the numerals for the following powers of 10:
   a) 10^6
   b) 10^8
   c) 10^{12}

We reviewed *exponents* so that we can use them to simplify our writing of numerals in *expanded form.*

Now that we see that 1000 can be written as 10^3 and 100,000 can be written as 10^5, we can substitute the shorter form.

EXAMPLE: Write 4795 in *expanded form, using exponents.* We have already seen that 4795 = (4 x 1000) + (7 x 100) + (9 x 10) + (5 x 1).

Using *exponents*, we would write this:
$4795 = (4 \times 10^3) + (7 \times 10^2) + (9 \times 10^1) + (5 \times 1)$

Note: (5 x 1) can also be written (5×10^0).

If you do not think this is a simpler method, let's try a number with more digits.

EXAMPLE: Write 2,765,931 in *expanded form.*

$$2{,}765{,}931 = (2 \times 1{,}000{,}000) + (7 \times 100{,}000) + (6 \times 10{,}000) + (5 \times 1000) + (9 \times 100) + (3 \times 10) + (1 \times 1)$$

Substituting powers of 10:

$$2{,}765{,}931 = (2 \times 10^6) + (7 \times 10^5) + (6 \times 10^4) + (5 \times 10^3) + (9 \times 10^2) + (3 \times 10^1) + (1 \times 10^0)$$

NOTE: *IF A PLACE VALUE IS MISSING WHEN A NUMBER IS WRITTEN IN EXPANDED FORM, THEN THE DIGIT FOR THAT PLACE VALUE IS ZERO.*

EXERCISES:

1) Simplify each number in *expanded form* by using *powers of 10:*
   a) (3 x 1,000,000) + (5 x 100,000) + (6 x 1000) + (2 x 100) + (1 x 10) + (7 x 1)
   b) (9 x 100,000) + (4 x 10,000) + (5 x 1000) + (7 x 100) + (2 x 1)

2) Write each numeral in expanded form:
   a) 70,342
   b) 96,530,421
3) Write the regular number represented by:
   $(8 \times 10^7) + (2 \times 10^6) + (5 \times 10^5) + (3 \times 10^4) + (9 \times 10^2) + (2 \times 10^1) + (6 \times 10^0)$

POSTTEST

1) Write a short definition: (no peeking!) *Exponent, base, place value, expanded form.*
2) Simplify, using exponents:
   a) 10 x 10 x 10 x 10 x 10 x 10
   b) 7 x 10,000
   c) 3 x 10 x 10 x 10 x 10
3) Write each of the following, using *both methods of expanded form:*
   a) 34,972 b) 7,634,125 c) 8,034,921
4) Write numerals for the following:
   a) $(3 \times 10^2) + (4 \times 10^1) + (9 \times 10^0)$
   b) $(6 \times 10^5) + (7 \times 10^4) + (9 \times 10^2) + (8 \times 10^1) + (1 \times 10^0)$
   c) $(4 \times 10^7) + (2 \times 10^6) + (5 \times 10^4) + (7 \times 10^3) + (2 \times 10^2) + (9 \times 10^0)$

OBJECTIVE 8-38

READING DECIMAL NUMERALS

OBJECTIVE:

The student will be able to read numbers containing decimal numerals up to the millionths place.

PRETEST

Read these numbers to your teacher:

1) 435,198.06	6) .99999
2) 23.4567	7) .110011
3) 143.9	8) .000548
4) 678.132	9) .236789
5) 1.000006	10) .00888

Let's review some of the things you have learned about the place values of *decimal fractions.* A *decimal fraction* is made up of the digits to the *right of the decimal point* in a number. If written as a proper fraction, the number would have a denominator of some power of 10. For example, 6.14 is read as "six and fourteen hundredths" and could be written 6 14/100.

You know that every whole number has a value *10 times greater* than the number to the *right* of it. The *tens place* is *10 times* greater than the *units (ones) place;* the *hundreds place* is *10 times* greater than the *tens place*, etc. You can see this relationship by comparing the powers of 10 that represent each place value:

Units	$= 10^0 =$	1
Tens	$= 10^1 =$	10
Hundreds	$= 10^2 =$	100
Thousands	$= 10^3 =$	1000

Each place to the *right of the decimal point* has the same relationship.

The *first place to the right of the decimal point is called the tenths place.* (Notice that all place values to the right of the decimal point end in *-ths.*)

.1 is read "one tenth"
.5 is read "five tenths"
.8 is read "eight tenths"

Read these numbers to yourself: .2 .9 .6 .3

The *units place* is *10 times greater* than the *tenths place.* The *tenths place* is 1/10 of the *units place.*

<u>STUDY THIS PLACE VALUE TABLE:</u>

10^9	1,000,000,000	1,	Billions	10^9
10^8	100,000,000	8	Hundred-Millions	10^8
10^7	10,000,000	9	Ten-Millions	10^7
10^6	1,000,000	5,	Millions	10^6
10^5	100,000	2	Hundred-Thousands	10^5
10^4	10,000	6	Ten-Thousands	10^4
10^3	1,000	3,	Thousands	10^3
10^2	100	9	Hundreds	10^2
10^1	10	7	Tens	10^1
10^0	1	4	Units (Ones)	10^0
10^{-1}	.1	.1	Tenths	$\frac{1}{10^1}$
10^{-2}	.01	3	Hundredths	$\frac{1}{10^2}$
10^{-3}	.001	2	Thousandths	$\frac{1}{10^3}$
10^{-4}	.0001	4	Ten-Thousandths	$\frac{1}{10^4}$
10^{-5}	.00001	5	Hundred-Thousandths	$\frac{1}{10^5}$

Notice that decimal fractions can be written as *negative powers of 10* or as fractions with a positive power of 10 in the denominator.

When there are *2 digits* to the *right of the decimal point*, the digit in the *second place* is said to be in the *hundredths place.* (Notice the spelling, again.)

.03 is read "three hundredths"
.01 is read "one hundredth"
.17 is read "seventeen hundredths"

Read these numbers to yourself: .05 .07 .9 .02 .55

When there are *3 digits* to the *right of the decimal point*, the digit in the *third place* is said to be in the *thousandths place.* (Notice spelling.)

.001 is read "one thousandth"
.035 is read "thirty-five thousandths"
.128 is read "one hundred twenty-eight thousandths"

Read these numbers to yourself:

.159 .608 .72 .7 .701
.49 .884 .11 .23 .55

The *hundredths place* is *10 times greater than the thousandths place. The thousandths place is 1/10 of the hundredths place.*

When there are *4 digits* to the *right of the decimal point*, the digit in the *fourth place* is said to be in the *ten-thousandths place.* (Notice spelling.)

.0018 is read "eighteen ten-thousandths"
.1667 is read "one thousand six hundred sixty-seven ten-thousandths"
.1075 is read "one thousand seventy-five ten-thousandths"

Read these numbers to yourself:

.0179 .08 .2184 .5783
.8808 .008 .7965 .103

The *thousandths place* is *10 times greater than the ten-thousandths place.* The *ten-thousandths place* is 1/10 of the *thousandths place.*

When there are *5 digits* to the *right of the decimal place*, the digit in the *fifth place* is said to be in the *hundred-thousandths* place.

.12345 is read "twelve thousand three hundred forty-five hundred-thousandths"
.08908 is read "eight thousand nine hundred eight hundred-thousandths"

Read these numbers to yourself:

.06981	.0816	.01976
.052	.76543	.33
.10101	.897	.23454
.07	.76321	.08016

The *ten-thousandths* place is *10 times greater* than the *hundred-thousandths* place. The *hundred-thousandths* place is 1/10 of the *ten-thousandths* place.

NOTICE: THERE ARE NO COMMAS IN DECIMAL FRACTIONS

Did you also notice that you read a decimal fraction just like a whole number and name it by the place value of the last digit to the right of the decimal point???

EXERCISE:

Write these mixed numbers. They contain a whole number and a decimal fraction. Remember that "and" tells you where to place the decimal point. The *only* time you can say "and" when reading a number is when you come to the decimal point.

1) fifty-six *and* sixty-two hundredths
2) one thousand eight hundred eight *and* eight hundred eight ten-thousandths
3) five hundred ninety *and* nine thousandths

Read these numbers to the teacher:
37.6 49.909 376.0367 3.03003 198.03
5,876,102.85 36,987.0543 47,600.0062

CHALLENGE! Read this number correctly:

2,006,598,362.012349

When there are *6 digits* to the *right of the decimal point*, the digit in the *sixth place* is said to be in the millionths place.

.698504 is read "six hundred ninety-eight thousand, five hundred four millionths"

.156392 is read "one hundred fifty-six thousand three hundred ninety-two millionths"

.000008 is read "eight millionths"

Read these numbers to yourself:

.592368	.09634	.000001
.764	.0663	.88
.101110	.554392	.00907

The *hundred-thousandths place* is 10 times greater than the *millionths place.*

The *millionths place* is 1/10 of the *hundred-thousandths place.*

POSTTEST

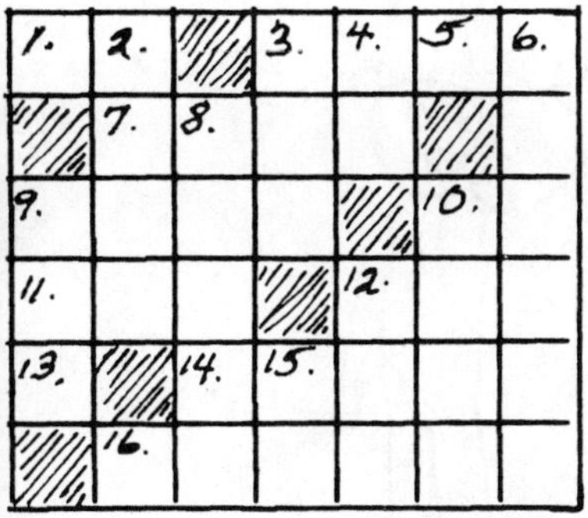

ACROSS:

1. eight hundredths
3. one hundred fifty-seven and six tenths
7. seven thousand two hundred fifty-one ten-thousandths
9. one thousand four hundred eighty-six ten-thousandths
10. twenty-five hundredths
11. ninety-three and six tenths
12. two hundred fifteen thousandths
13. eight
14. five and five thousand five hundred eighty-eight ten-thousandths
16. three hundred twenty-four thousand three hundred ninety-six millionths

DOWN:

1. no tenths
2. eight and seven hundred forty-three thousandths
3. one hundred fifty-six
4. fifty-one
5. seven
6. six hundred fifteen thousand five hundred eight-six millionths
8. twenty-eight and six hundred fifty-two thousandths
9. one hundred ninety-eight thousandths
10. two thousand one hundred eighty-nine ten-thousandths
12. two hundred fifty-three thousandths
15. fifty-four hundredths
16. three tenths

OBJECTIVE 8-40

MULTIPLYING DECIMALS

<u>OBJECTIVE:</u>

The student will be able to multiply decimal numerals up to the millionths place.

<u>PRETEST</u>

Answers are given for each product. Correctly place the decimal point in each product.

1) 1.21 x 0.6:	726
2) .3 x 53.8;	1614
3) 2.67 x 1.5;	4005
4) 71.3 x 2.01;	143313
5) 2.5 x 0.011;	275

Multiply:

1) 3.9 x 6.91003	6) 2.00 x 3.717
2) 4.32 x 7.3	7) 3.652901 x 17.6
3) 5.05 x 2.178	8) 3.754 x 7.563
4) 5.77 x 9.221304	9) 11.12573 x 13.16
5) 17.38 x 6.367	10) 12 x 7.8116

The following examples show how fractions help to decide where the decimal point should be placed in the product:

<u>EXAMPLE:</u> $.6 \times .9 = \frac{6}{10} \times \frac{9}{10} = \frac{54}{100}$ <u>OR</u> $.54$

<u>EXAMPLE:</u> $.4 \times .08 = \frac{4}{10} \times \frac{8}{100} = \frac{32}{1000}$ <u>OR</u> $.032$

STUDY THE FOLLOWING EXAMPLES. DO YOU SEE THE PATTERN?

1.3 x 1.5 = 1.95	.13 x 1.5 = .195
1.4 x 1.4 = 1.96	.14 x 1.4 = .196
1.3 x 1.7 = 2.21	.13 x 1.7 = .221

Have you discovered that the number of decimal places in the product equals the number of places in the multiplicand and the number of places in the multiplier?

Notice:

1.3	(multiplicand)	= 1 decimal place
x 1.5	(multiplier)	= 1 decimal place
1.95	(product)	= 2 decimal places

If the product ends in one or more zeros, the zeros are counted as places when deciding where to put the decimal point.

EXAMPLE:

.20	(2 decimal places)
x 3.0	(1 decimal place
.600	(3 decimal places)

From these examples, you can see that the placing of the decimal point can be determined by *adding together* the number of digits in the multiplicand and the number of digits in the multiplier that are *to the right of the decimal points.*

IF THE NUMBER OF DIGITS IN THE PRODUCT IS LESS THAN THE NUMBER OF PLACES REQUIRED, ADD ENOUGH ZEROS TO GET THE REQUIRED NUMBER OF PLACES. ADD THE ZEROS IMMEDIATELY TO THE RIGHT OF THE DECIMAL POINT AND BEFORE THE DIGITS IN THE PRODUCT.

EXAMPLE:

.05	(2 decimal places)
x .3	(1 decimal place)
.015	(3 decimal places)

EXAMPLE:

```
   .13  (2 decimal places)
  x.05  (2 decimal places)
 .0065  (4 decimal places)
```

POSTTEST

1) How many decimal places will there be in each product?

a) .2 x .3 c) .003 x 2 e) .02 x .03

b) .2 x 3 d) .02 x .3 f) 2 x 3

2) The decimal point has been left out of these products. Copy the product and place the decimal point in the proper place.

a) .6 x 4 = 24
b) 39 x .16 = 624
c) 1.9 x 6 = 114
d) 7.5 x 84 = 6300
e) .37 x .013 = 481
f) .42 x .003 = 126

3) Find the products: (Apply what you have learned in this packet to help you determine where the decimal point goes. The same rule applies no matter how many digits there are to the right of the decimal point.)

a) 1.7 x 2.9
b) 16 x 43.8
c) 5.16 x .03001
d) 17.8 x .006
e) 0.15 x 0.153
f) 2.518 x 1.9
g) 6.72 x .25
h) 2.4 x 10.1075
i) .171 x 5.182003
j) .002 x .0024
k) 5000 x 32.18
l) 134.7 x 1.015423

OBJECTIVE 8-41

DIVISION OF DECIMALS

OBJECTIVE:

The student will be able to divide decimal numerals using divisors up to the ten-thousandths place.

PRETEST

DIVIDE

1) $.2\overline{)\,.4}$

2) $.004\overline{)\,3.6}$

3) $.0009\overline{)\,63}$

4) $.41\overline{)\,.246}$

5) $1.4\overline{)\,29.4}$

6) $.0096\overline{)\,576}$

7) $9.2\overline{)\,8096}$

8) $1.3\overline{)\,7.839}$

9) $.09\overline{)\,11799}$

10) $1.0403\overline{)\,10.403}$

To understand the division of decimals, we will first have to review the process of multiplication by powers of 10. It is really very simple: move the point in the number to the *right* as many places as there are zeros in the power of 10. To multiply by 10, move the decimal point *one place* to the *right;* by 100, *two places* to the right; by 1000, *three places* to the *right*, etc.

EXAMPLES:

2.5678 x 10 = 25.678
2.5678 x 100 = 256.78
2.5678 x 1000 = 2567.8
2.5678 x 10,000 = 25,678
Can you see that in order to change 2.5678 to a whole number, you must multiply by 10,000?

When using decimal numerals in division, we must first make sure that the *divisor* (number we are dividing by) is a *whole number.*

This will be easier to understand, if we consider the following facts:

$.61\overline{)\,109.8}$ can be written in the form of a fraction:

$$\frac{109.8}{.61}$$

We also know that we can find an equivalent fraction by multiplying *both the numerator and the denominator by the same number.*

> To make a whole number out of .61, we must multiply it by 100; if we multiply .61 by 100, we must also multiply 109.8 by 100.
> Our new *equivalent* fraction is: $\frac{10980}{61}$
> Notice that we had to add a zero in the dividend, because there were not enough places.

If the divisor is already a whole number, then all we have to do is place the decimal point in the *quotient* (answer) *directly over the decimal point in the dividend.*

EXAMPLE: $8\overline{)88.08}$

8 is already a whole number, so do not change it.

$$\begin{array}{r} 11.01 \\ 8\overline{)88.08} \end{array}$$

EXERCISE:

Multiply each of the following by the correct power of 10 to make it a whole number. (Remember that $10^0 = 1$.)

Write the example out: $23.45 \times 10^2 = 2345.$

1) 6.23
2) .058
3) 29.6
4) 145,089
5) .07
6) 17.2
7) .005
8) .033
9) .0102
10) 3.0204

Let's go over the rules for division of decimal numerals once more:

Step 1:

a) If the *divisor* is already a whole number, place the decimal point in the *quotient* directly over the decimal point in the *dividend.*

b) If the *divisor* is not already a whole number, multiply by the power of 10 necessary to mak it a whole number.

Step 2:

Multiply the *dividend* by the same power of 10.

Step 3:

Place the decimal point in the *quotient* directly above the new position of the decimal point in the *dividend.*

EXAMPLE $28\overline{)15.4}$

(In this example, we will have to add a zero to the right of the decimal point in the dividend in order to finish the division. Otherwise, we would have a *remainder.* We can do this, because adding zeros to the *right* of the point does not change the value of the decimal fraction.

.4 = 4/10; .40 = 40/100; and 40/100 = 4/10

```
      .55
28 | 15.40
     14 0
      140
      140
```

EXERCISE: DIVIDE:

1) $2.3\overline{)1.38}$
2) $23\overline{)1.38}$
3) $.023\overline{).0138}$
4) $.0023\overline{)13.8}$
5) $.0023\overline{)138}$
6) $230\overline{)13.8}$

POSTTEST

1) Define these words: *divisor*, *remainder*, *quotient*, *dividend.*
2) Find one-half of .0001 by division.
3) How many places would you move the decimal point when multiplying by these powers of 10?
   a) 10^0
   b) 10^3
   c) 10^1
   d) 10^{10}
   e) 10^{15}
4) *Multiply:*

By 10^0	*By* 10^1	*By* 10^2	*By* 10^3
a) 25	d) 14.89	g) 35	j) 300
b) 2.5	e) .005	h) 20.9	k) 2.487
c) .025	f) 33	i) .02056	l) .009

5) If candy bars cost $1.65 for a bag of 30, how much is each bar?
6) If a telephone call costs $.30 for the first three minutes and $.05 for each additional minute, how long was the call for which the charges were $1.05?
7) *Divide:*
   a) $.04 \overline{)76}$
   b) $3.65 \overline{)208.05}$
   c) $.0003 \overline{)2.799}$
   d) $4.007 \overline{)52.091}$
   e) $1.034 \overline{)930.6}$

OBJECTIVE 8-66

SCIENTIFIC NOTATION

OBJECTIVE:

The student will be able to use large and small numbers in scientific notation.

PRETEST

Change the following numbers into a simpler form, using scientific notation:

1) 93,000,000
2) 1570
3) 12,156,000
4) 100,025
5) 15,600.15
6) 100.15
7) 19,145
8) 201.007

Sometimes, numbers are too large to work with easily, so a method of rewriting large numbers has been developed. It is called "*scientific notation*," because scientists usually write very large or very small numbers in this way.

Let's see how it works.

930	can	be	rewritten	as	93 x 10 or 93 x 10^1
5600	"	"	"	"	56 x 100 or 56 x 10^2
54,500	"	"	"	"	545 x 100 or 545 x 10^2
720,000	"	"	"	"	72 x 10,000 or 72 x 10^4

Now we need one more step to write these numbers in the form of *scientific notation. The number being multiplied by the power of 10* must *be a* number between 0 and 10. In other words, it must be some number 1 – 9.

Let's consider the above numbers.

93 is *not* a number between 0 and 10, so we must make it one by placing a decimal point between the 9 and the 3. The number becomes 9.3. Now, in order to change 9.3 to 930, we must multiply by 10^2, so 930 written in *scientific notation* is: 9.3×10^2.

NOTE: TO MULTIPLY BY POWERS OF 10, MOVE THE DECIMAL POINT TO THE RIGHT. TO DIVIDE BY POWERS OF 10, MOVE THE DECIMAL POINT TO THE LEFT. THE EXPONENT TELLS YOU HOW MANY PLACES TO MOVE THE POINT IN EITHER DIRECTION.

$56 \times 10^2 = 5.6 \times 10^3$ *in scientific notation.*
Check: $56 \times 10^2 = 56 \times 100 = 5600$ is the same as
$5.6 \times 10^3 = 5.6 \times 1000 = 5600$

$545 \times 10^2 = 5.45 \times 10^4$ in *scientific notation.*
Check: $545 \times 10^2 = 545 \times 100 = 54{,}500$ is the same as
$5.45 \times 10^4 = 5.45 \times 10{,}000 = 54{,}500$

$72 \times 10^4 = 7.2 \times 10^5$ *in scientific notation.*
Check: $72 \times 10^4 = 72 \times 10{,}000 = 720{,}000$ is the same as
$7.2 \times 10^5 = 7.2 \times 100{,}000 = 720{,}000$

Let's go over the steps again:

Step 1: Place the point so that the whole number is a number between 0 and 10.

Step 2: To find the exponent that you will use as the power of 10, count the number of places to the right of the point that you need to get to the original number.

EXAMPLE: Change 19.683 to *scientific notation.*

Step 1: 1.9683

Step 2: To get to 19.683 from 1.9683, you must move *one* place to the right. The exponent of 10 is 1, so
19.683 written in scientific notation is
1.9683×10^1

EXERCISE: Write these numbers in scientific notation:

1) 15,600,000,000
2) 9,875,000,000,000
3) 1,007,000,000,000
4) 156.817
5) 31,487,500,000,000

Now that you have learned how to write *very large* numbers, we will try some *very small* numbers. We will follow the same steps except that we will be dealing with *negative exponents. A negative exponent* represents a *fraction* with a power of 10 in the *denominator.*

STUDY THIS TABLE:

10^{-1} = .1 = 1/10
10^{-2} = .01 = 1/100
10^{-3} = .001 = 1/1000
10^{-4} = .0001 = 1/10,000

Can you see that multiplying by a power of 10 with a *negative exponent* is the same as multiplying by a fraction? Also, when a fraction has a 1 in the numerator (a *unit fraction*), all you have to do is divide by the denominator. So, when the exponent is *negative, divide* by the denominator of the fraction represented by the exponent. When the exponent is −1, divide by 10; −2, divide by 100; −3 divide by 1000; −4, divide by 10,000; −5, divide by 100,000; −6, divide by 1,000,000, etc. Remember that when you divide by powers of 10, move the decimal point the required number of places to the *left.*

CHALLENGE! You have just discovered that when you *multiply* by *negative* powers of 10, you move the point to the *left* instead of to the right as you do for *positive* powers of 10. Explain why you would move the decimal point to the *right*, if you were *dividing by negative* powers of 10. (Hint: Make use of the fact that positive and negative numbers are opposites: −5 is the opposite of +5, etc.)

<u>EXAMPLE:</u> Write .0025 in *scientific notation.*

<u>Step 1:</u> Place the decimal point correctly: 2.5

<u>Step 2:</u> Count the number of places you have to go *left* to get from 2.5 back to .0025. You have to go 3 places. Therefore, your exponent is −3, so

$.0025 = 2.5 \times 10^{-3}$ in *scientific notation.*

Check: $2.5 \times 10^{-3} = 2.5 \times 1/1000 =$

$$\frac{2.5}{1000} = \frac{25}{10{,}000} = .0025$$

OR $2.5 \times .001 = .0025$

<u>EXAMPLE:</u> Write .3217845 *in scientific notation.*

<u>Step 1:</u> 3.217845

<u>Step 2:</u> To get from 3.217845 to .3217845, move the decimal point *one place* to the left. This means you are dividing by 10, so your exponent will be −1.

$.3217845 = 3.217845 \times 10^{-1}$

NOTICE THAT NO MATTER HOW MANY NUMBERS ARE INVOLVED, THE IMPORTANT THING TO REMEMBER IS THE MOVEMENT OF THE DECIMAL POINT.

<u>EXERCISE:</u> *Write the following numbers in* <u>scientific</u> *notation:*

1) .0033
2) .198
3) .0007
4) .123456
5) .00050005

POSTTEST

1) It would take 3,000, 000, 000, 000, 000, 000, 000, 000, 000 candles to give as much light as the sun. Write this number in scientific notation.
2) The number of grains of sand on the beach in Malibu is 1.0×10^{14}. Write this number in long form.
3) The weight of the water in all of the oceans of the world is 1,580,000,000,000,000,000 tons. Write this number in scientific notation.
4) A *googol* is the number named by 1 followed by 100 zeros. Write this number correctly in long form, being sure to get the commas in the right places. Then, write it in scientific notation.
5) Write these large numbers in scientific notation:
   a) 7,123,000,000,000
   b) 6,000,000,000
   c) 13,000,000,000,000
   d) 435,008,011
   e) 10,000,000,225
6) Write these small numbers in scientific notation:
   a) .7
   b) .0011
   c) 10.0000000002
   d) .890023675
   e) .000000000000039816

OBJECTIVE 8-91
SOLVING SIMPLE EQUATIONS

OBJECTIVE: The student will be able to solve simple algebraic equations involving the sum or difference of two terms.

PRETEST

1. What equation in the following set is *not equivalent* to *d?*

   a) $3a + 2 - 2 = 14 - 2$ c) $3a + 2 + 5 = 14 + 5$
   b) $3a + 2 + 4 = 14 + 4\frac{1}{2}$ d) $3a = 12$

2. What is the value of *a* in the equivalent equations?
3. What is an *equation?*
4. What is a *variable?*

An equation is an open number sentence which states that two quantities are equal. The quantity on the left side of the = sign is called the *left member;* the quantity on the right side of the = sign is called the *right member.* State the *right member* and the *left member* in the following equations:

$$6x = 3x + 9$$
$$4p - 2p = 56$$
$$5b - 6 = 4b + 3$$

In the above equations, x, p, and b are called *variables.* A *variable* is any symbol used to represent a number we do not know. When we find out the number that the variable is representing, we can replace the variable by that number. If our work is correct, we will have a statement that both sides of the = sign are the same.

For example, we will replace the b in the third equation by 9. Do you recall that if there is no sign of operation between numbers that means to multiply the two numbers? 5(6) means multiply 5 times 6; likewise, 5b means multiply 5 times b. All right, let's replace each b in the equation by 9. It will look like this: $5(9) - 6 = 4(9) + 3$. Then, $45 - 6 = 36 + 3$ and $39 = 39$.

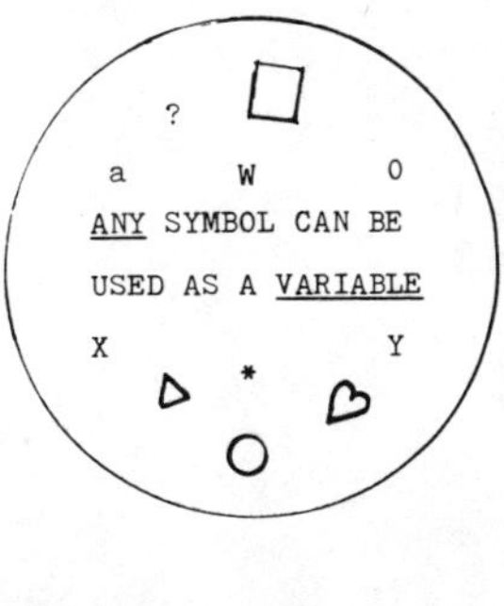

Now, you are going to learn how to find the number represented by the variable in an equation. We have an axiom (rule) which we call the *Axiom of Opposites.* This Axiom states that if you add a number and its *opposite,* the result will be zero. It is written like this: $a + (-a) = 0$. a is the number; $-a$ is the *opposite* of a.

Do you see that to get the opposite of a number all you have to do is change the sign of the number? If the number is positive (+5), its opposite is negative (−5); if the number is negative (−15), its opposite is positive (+15). If the number has *no sign* in front of it, it is considered to be *positive. Name the opposites of the following numbers:*

35 +109 −32 −¼ +.25 −1½ 75

Now that you know how to find the opposite of any number, you will be able to apply the *Axiom of Opposites* to equation solving. We will solve this equation: $2x + 4 = 8$

$2x + 4 - 4 = 8 - 4$ (To get the 4 over with the 8 on the right side of the equal sign, add the *opposite* of 4 to *both sides* of the equation.)

Since $4 - 4 = 0$ and $8 - 4 = 4$, our equation now reads: $2x = 4$. Now, divide both sides of the equation by 2 to find the value of x. $x = 2$. To prove that $x = 2$, substitute 2 for x in the original equation:

$$2x + 4 = 8$$
$$2(2) + 4 = 8$$
$$4 + 4 = 8$$
$$8 = 8$$

Now, let's try an equation with subtraction:

$$2x - 4 = 8$$
$$2x - 4 + 4 = 8 + 4$$
$$2x = 12$$
$$x = 6$$

EXERCISE:

(Remember that answers do not necessarily have to be whole numbers.)

1. $6x + 3 = 21$
2. $4a + 2 = 26$
3. $a - 5 = 3\frac{1}{2}$
4. $10p - 2 = 88$
5. $7s - 15 = 20$
6. $10 + 3c = 43$

CHALLENGE! $3a + 5a + 8 - 10 = a + 40$

POSTTEST

1) Define in your own words: *equation, variable, left member, right member, Axiom of Opposites.*
2) Find the *solution set* (answers) for the following equations:

a) $2c + 3 = 9$
b) $6x - 6 = 0$
c) $47 = 3 - 2 + x$
d) $3c + 18 - 15 = 7 - 3$
e) $2y - 10 = 36$
f) $9c - 9 = 27$
g) $a + 16 = 0$
h) $H - 2\frac{1}{2} = 2\frac{1}{2}$
i) $3\ 1/8 + ? = 4\frac{1}{4}$
j) $\overset{*}{\underset{*}{}} + .02 = .9$

OBJECTIVE 8-114

METRIC MEASURE OF LIQUIDS AND WEIGHT

OBJECTIVE:

The student will be able to define and use the various units of liquid and weight measure in the Metric System.

PRETEST

1) Define the prefixes used in metric measure:
   a) centi- b) milli- c) deci- d) kilo-
2) If you were mailing a present that weighed 275 grams to a friend, what would be the weight in kilograms?
3) If the gas tank in a car holds 16 gallons, how many liters would it hold?
4) A *kilogram* = ________ grams.
5) A *liter* is equal to about ________ liquid quarts.

The unit of measure for liquids in the metric system is the *liter.* The *liter* is equivalent to about *1.1 liquid quarts* and the quart is *.95 of a liter.* You can see, then, that a quart is *smaller* than a *liter.* When our country has changed over completely to the metric system of measure, people will not have to worry about changing English units (the measures we use now) to metric units. This will be a long, hard process, so you should become acquainted with the way units are *converted* from one system to the other.

Can you see that a gallon would be *less* than *4 liters?* A gallon would be *3.8 liters.* Since there are 4 quarts in a gallon and a quart is equal to .95 liter, multiply 4 x .95 to get the number of *liters in a gallon.*

<u>EXERCISE</u>:

1) If the people in your family drink 3½ liters of milk every day, how many quarts do they drink in a week?
2) Change 10 quarts to liters.

All you need to remember is that when you want to change *liters to quarts,* multiply the number of liters by 1.1; *quarts to liters,* multiply the number of quarts by .95.

Just as the *meter* is the *unit of measure* for linear measurement, the *liter* is the unit of measure for liquid measurement. Therefore, we can use the same prefixes to denote measures that are larger or smaller than the *liter:*

A *deciliter* = 1/10 of a liter (dl.)
A *centiliter* = 1/100 of a liter (cl.)
A *milliliter* = 1/1000 of a liter (ml.)
A *kiloliter* = 1000 liters (kl.)

So, you can see that:

A liter = 10 deciliters
100 centiliters
1000 milliliters

The *unit of weight* in the Metric System is the *gram*. A gram is a *very* small weight—only about .035 ounce.

The *kilogram* is used to weigh things that we usually weigh in pounds. A *kilogram* is equal to *1000 grams* and weighs about 2.2 pounds. If you weigh 110 pounds, then your weight in *kilograms* would be *50 kilograms.* (Since a *kilogram = 2.2 pounds*, divide 110 by 2.2). People should be very pleased with this aspect of the Metric System, because they will seem to weigh less. Also, everyone will probably have a hard time, at first, figuring this out. Since you are doing this LAP, you will be one step ahead of others who are not making any effort to become knowledgeable about the Metric System. Too many people seem to think that if they ignore the Metric System, it will go away. They probably do not realize the extent to which it is already in use and the plans that are being made by our government to get it into operation. If you look around closely, you will be surprised to see how many measurements are already in metric.

We can also use the prefixes to denote measures that are larger or smaller than the *gram:*

A *decigram* = 1/10 of a gram (dc.)
A *centigram* = 1/100 of a gram (cg.)
A *milligram* = 1/1000 of a gram (mg.)
A *kilogram* = 1000 grams (kg.)

So you can see that:

A gram = 10 *decigrams*
100 centigrams
1000 *milligrams*

Another measurement that it might be useful to know is that a metric ton (M.T.) weighs *1000 kilograms.*

EXERCISE:

1) 1832 kilograms = ____________ grams.
2) Chemists often weigh small objects to the nearest ten-thousandths of a gram. What decimal part of an ounce is that? (Hint: 1 gram = .035 ounce)
3) Mary weighs 150 pounds. For her size and weight, she should weight 59,090 grams. How many pounds overweight is she?

POSTTEST

1) A Datsun pick-up truck weighs ¾ of a ton (1 ton = 2000 pounds). How much does it weigh in kilograms?
2) If an automobile gets 18 miles to the gallon of gas, how many liters of gas will it burn in 75 miles?
3) A person's weight on earth is approximately 6 times that on the moon. If he weighs 157 pounds on earth, what would his weight be in kilograms on the moon?
4) The African elephant can weigh as much as 5 tons (10,000 lbs.) What would this weight be equal to in kilograms?
5) Jane's mother bought ½ liter of imported olive oil to use in making salads and for cooking. Is that more or less than two quarts?
6) If you are allowed to take 40 pounds of luggage on an airplane flight, how many kilograms does it weigh?

OBJECTIVE 8-118
ROUNDING OFF DECIMALS

OBJECTIVE:

The student will be able to round off decimal numerals to any given place value.

PRETEST

Round off to the given place value:

Nearest tenth:	*Nearest hundredth:*	*Nearest thousandth:*
1) .68	3) 8.349	5) 1.9685
2) 5.425	4) .5146	6) 3.24728

Nearest unit:	*Nearest ten:*	*Nearest millionth:*	*Nearest thousand:*
7) 57.18	9) 146.22	11) .0089347	13) 45,683
8) 89.562	10) 89.1	12) 2.0500302	14) 9,123,468

We "round off" a number to express it approximately or to a given number of decimal places. When numbers are very large or very small, it is easier to work with them if they are rounded off. It is easier to *compare* rounded numbers.

You have seen statements like this in your science books: "The sun is about 93,000,000 miles from Earth. It has a diameter of approximately 864,000 miles and a mass about 330,000 times that of Earth." Scientists do not know the *exact* figures, so they use rounded numbers.

In making graphs, large numbers are usually rounded off.

When finding decimal equivalents for fractions, we often have to round off, because the *quotient* either *repeats or does not terminate.*

<u>EXAMPLE:</u> 1/3 = .33333. . ., so we round it off to .3
1/7 = .142857142857. . ., so it could be rounded off as: .14; .143; .1429; .14286, etc. It would probably be rounded off to .14, since the numbers to the right are so small that they would not affect calculations very much.

Now, let's see how we go about rounding off a number:

<u>EXAMPLE:</u> Round off 98.92 to the *nearest unit.*

<u>Step 1:</u> *Find the digit* in the *units (ones) place.* The digit is 8.

<u>Step 2:</u> Look at the digit to the *right* of the units place. The digit is 9.

Step 3: Determine whether the digit to the right of the units place is 5, larger than 5, or smaller than 5. Since the digit is 9, it is *larger than 5.*

Step 4: When the digit to the right of the place is larger than 5, we "round up," that is, we change the digit in the units place to the next higher digit. Thus, the 8 becomes 9.

Step 5: Drop all digits to the right of the digit being rounded off. 98.92 rounded off to the nearest unit is 99.

EXAMPLE: Round off 1,086,343 to the nearest hundred thousand.

Step 1: The digit in the hundred thousands place is 0.

Step 2: The digit to the right of 0 is 8.

Step 3: 8 is larger than 5, so round up. 0 becomes 1.

Step 4: Change all of the digits to the right of the place being rounded off to zeros. We have to do this, because none of them are to the right of the decimal point. We need them as place-holders, so we will not change the value of the number. 1,086,343 rounded off to the nearest hundred thousand is 1,100,000.

EXAMPLE: Round off 3.45 to the nearest tenth.

Step 1: The digit in the tenths place is 4.

Step 2: The digit to the right of 4 is 5.

Step 3: When the digit to the right of the digit being rounded is 5, you can either *round up or round down* (leave the number the way it is and drop or change to zero all the digits to the right of the digit being rounded.)
4 becomes 5

3.45 rounded off to the nearest tenth is 3.5

EXAMPLE: Round off .0286934 to the nearest millionth.

Step 1: *The digit in the millionths place is 3.*

Step 2: The digit to the right of 3 is 4.

Step 3: 4 is less than 5, so *round down* -leave the number the way it is.

Step 4: Drop the 4 in the ten millionths place.

.0286934 rounded to the nearest millionth is .028693.

Let's review the rules for rounding off a number once more:

Step 1: Find the digit in the place to be rounded off.

Step 2: Determine whether the digit *to the right* of the place being rounded off is 5, less than 5 or greater than 5.

Step 3: If the digit to the right of the place being rounded off is 5 or greater than 5, round up to the next higher number. If it is less than 5, do not change the digit.

Step 4: Change all digits to the right of the place being rounded to zeros. If the zeros are in the decimal fraction part of a number, drop them.

EXERCISE: Round off to the given places:
1) 8726 to the nearest thousand.
2) 3,538,837 to the nearest ten thousand.
3) 1.0078 to the nearest thousandth.
4) .00754 to the nearest ten-thousandth.
5) 23.9 to the nearest unit.

POSTTEST

1) Round off 964,342,368,220 to the nearest million.
2) The average price of a home in Fall River is the highest in the state of Massachusetts, going from a price of $31,487 in 1976 to $43,349 in 1977. Give in round figures the amount that the price went up.
3) The price of gold is currently $145.70 per troy ounce. What would this price be to the nearest dollar?
4) If the price of 6 candy bars is 45¢, what will the price to the nearest cent of each bar be?
5) Do these division examples. Carry out the division two decimal places and round off the answer to the nearest tenth.

   a) $.96\overline{)8.8}$
   b) $7.5\overline{)456.2}$
   c) $2.2\overline{)3000}$
   d) $.7\overline{)9}$
   e) $25.4\overline{)83.35}$

APPENDIX I

ANSWER KEY

PRETESTS AND EXERCISES

PRETESTS—LEVEL 6

OBJECTIVES 6-(71,72)

Pretest:

1) None, 3, 5, 50, 3, 6
2) 10^1, 10^3, 10^0, 10^6, 10^4
3) 360; 5,000; .8; 167; 10.96
4) 15; 6.5; .2596; .0108
5) Move the decimal point *one* place to the *right* for each zero in the *multiplier*.
6) Move the decimal point *one* place to the *left* for each *zero* in the *divisor*.

OBJECTIVE 6-74

Pretest:

1) 2 squared *or* 2 to the second power.
   5 squared
   8 squared
   1.5 squared
2) 324 400 1024 210.25

Exercise:
25 13,924 250,000 1089 9801 576

Exercise:
184.96 475.24 10,060.09 3.3124 .0064

Exercise:
225 625 1225 3025 4225 5625 7225 9025

OBJECTIVE 6-79

Pretest:

1) 2
2) 1
3) 0
4) 15, 5, 3, 1
5) 11 and 1
6) If the last 3 digits of a number are divisible by 8, the whole number is divisible by 8.

Exercise:

1) Will vary
2) a. Yes
   b. No
   c. No
   d. Yes
   e. No
   f. Yes

Exercise:

a, c, d (5,3,17) are prime

Exercise:

Answers will vary.

Exercise:

a,c,d are divisible by 3

Exercise:

a, c, d, are divisible by 4

Exercise:

b, c, d, are divisible by 5

Exercise:

b, c, d are divisible by 6

Exercise:

b, c, d are divisible by 8

Exercise:

a, b, c are divisible by 9

Exercise:

b, d are divisible by 10.

OBJECTIVES 6-(80-83)

Pretest:

1. hundreds fours
2) $27_{10} = 11011_2$
3) $11_2 = 3_{10}$
4) one, zero, one, one, one Base 2

Exercise:

a) $10_2 = 2$
b) $1011_2 = 8 + 0 + 2 + 1 = 11$
c) $11_2 = 2 + 1 = 3$
d) $111_2 = 4 + 2 + 1 = 7$
e) $1001_2 = 8 + 0 + 0 + 1 = 9$
f) $101_2 = 4 + 0 + 1 = 5$
g) $1101_2 = 8 + 4 + 0 + 1 = 13$

1) 1_2
2) 10_2
3) 11_2
4) 100_2
5) 101_2
6) 110_2
7) 111_2
8) 1000_2
9) 1001_2
10) 1010_2

11) 1011_2
12) 1100_2
13) 1101_2
14) 1110_2
15) 1111_2
16) 10000_2
17) 10001_2
18) 10010_2
19) 10011_2
20) 10100_2

21) 10101_2
22) 10110_2
23) 10111_2
24) 11000_2
25) 11001_2
26) 11010_2
27) 11011_2
28) 11100_2
29) 11101_2
30) 11110_2

31) 11111_2
32) 100000_2
33) 100001_2
34) 100010_2
35) 100011_2
36) 100100_2
37) 100101_2
38) 100110_2
39) 100111_2
40) 101000_2

41) 101001_2
42) 101010_2
43) 101011_2
44) 101100_2
45) 101101_2
46) 101110_2
47) 101111_2
48) 110000_2
49) 110001_2
50) 110010_2

OBJECTIVE 6-84

Pretest:

1) a) 110111_2 b) 100111_2 c) 101011_2
2) a) 7 b) 19 c) 38
3)

32	16	8	4	2	1

4) a) 1010_2 b) 1101100_2
5) a) 10_2 b) 1000_2

Check: Answer shown in base 2 and 10.

$$\begin{array}{rcl} 11_2 & = & 2 + 1 = 3_{10} \\ \underline{10_2} & = & \underline{2} \quad + \underline{2_{10}} \\ 101_2 & & 4 + 1 = 5_{10} \end{array}$$

$$\begin{array}{rclcr} 111_2 & = & 4 + 2 + 1 & = & 7_{10} \\ \underline{+101_2} & = & \underline{4 + 0 + 1} & = & \underline{5_{10}} \\ 1100_2 & & 8 + 4 + 0 + 0 & = & 12_{10} \end{array}$$

Exercise:

a)

$$\begin{array}{rclcr} 10_2 & = & 2 + 0 & = & 2_{10} \\ 11_2 & = & 2 + 1 & = & 3_{10} \\ \underline{11_2} & = & \underline{2 + 1} & = & \underline{3_{10}} \\ 1000_2 & = & 8 + 0 + 0 + 0 & = & 8_{10} \end{array}$$

b)

$$\begin{array}{rclcr} 111_2 & = & 4 + 2 + 1 & = & 7_{10} \\ \underline{+10_2} & = & \underline{2 + 0} & = & \underline{2_{10}} \\ 1001_2 & = & 8 + 0 + 0 + 1 & = & 9_{10} \end{array}$$

c)

$$\begin{array}{rclcr} 1100_2 & = & 8 + 4 + 0 + 0 & = & 12_{10} \\ \underline{1011_2} & = & \underline{8 + 0 + 2 + 1} & = & \underline{11_{10}} \\ 10111_2 & = & 16 + 0 + 4 + 2 + 1 & = & 23_{10} \end{array}$$

d)

$$\begin{array}{rcl} 1110_2 & = 8+4+2+0 & = 14_{10} \\ \underline{1011_2} & = \underline{8+0+2+1} & = \underline{11_{10}} \\ 11001_2 & =16+8+0+0+1 & = 25_{10} \end{array}$$

e)

$$\begin{array}{rcl} 10110_2 & = 16+0+4+2+0 & = 22_2 \\ 1111_2 & = 8+4+2+1 & = 15_{10} \\ \underline{10101_2} & = \underline{16+0+4+0+1} & = \underline{21_{10}} \\ 111010_2 & = 32+16+8+0+2+0 & = 58_{10} \end{array}$$

Practice:

a)

$$\begin{array}{rcl} 1110_2 & = 8+4+2+0 & = 14_{10} \\ \underline{-101_2} & = \underline{4+0+1} & = \underline{-5_{10}} \\ 1001_2 & 8+0+0+1 & = 9_{10} \end{array}$$

b) 1110_2 c) 10010_2 d) 111_2 e) 100_2

OBJECTIVE 6-85

Pretest:

a) 110_2 b) 100111_2 c) 11001011_2
d) 1000110_2 e) 1010001_2

2. a) 1001_2 R = 100_2 b) 11_2 R = 10_2

Partial Products:

a)

$$\begin{array}{r} 1\,1\,1\,0_2 \\ \underline{\times\ 1\,0\,1_2} \\ 1\,1\,1\,0 \\ 0\,0\,0\,0 \\ \underline{1\,1\,1\,0} \\ 1\,0\,0\,0\,1\,1\,0_2 \end{array}$$

b)

$$\begin{array}{r} 1\,1\,1_2 \\ \times\ 1\,0\,1_2 \\ \hline 1\,1\,1 \\ 0\,0\,0 \\ 1\,1\,1 \\ \hline 1\,0\,0\,0\,1\,1_2 \end{array}$$

c)

$$\begin{array}{r} 1\,1\,0\,1_2 \\ \times\ 1\,1\,0_2 \\ \hline 0\,0\,0\,0 \\ 1\,1\,0\,1 \\ 1\,1\,0\,1 \\ \hline 1\,0\,0\,1\,1\,1\,0_2 \end{array}$$

a) 1000110_2 b) 100011_2 c) 1001110_2

Exercise:

a) 111000_2
b) 10101000_2
c) 1010000_2
d) 110001_2

Exercise:

a) 101_2 R = 100_2
b) 1001_2 R = 100_2
c) 11_2 R = 10_2

OBJECTIVES 6-(86-88)

Pretest:

1. a) one tenth b) one hundredth c) one thousandth d) ten e) one hundred f) one thousand
2. meter: 3.37 inches
3. a) 10^{-1} b) 10^{-2} c) 10^{-3} d) 10^{1} e) 10^{2} f) 10^{3}

Answers to Questions in Illustration:

100, 100:1000, a penny is 1/100 of a dollar, 10 years

Exercise:
1) 10
2) 100
3) 10
4) 100
5) 10
6) 1000

Exercise:
A kilometer measures 3280.8 feet. There are 5,280 feet in a mile, so you would travel less than 90 miles to get to Paris. (A kilometer is .621 miles or approx. 5/8 mile.)

Exercise:
1. decameter, hectometer, kilometer
2. kilometer
3. kilometer
4. 50

Challenge:
A kilometer = approx. 5/8 mi. so a mile is 8/5 kilometers.

30 x 8/5 = 48 kilometers.

OBJECTIVES 6-(91,92)

Pretest:
1) b, c, d, e
2) a) 5/100 b) 8/10 c) 9/1000 d) 7/100
3) a) 50% b) 25% c) 20%
4) A statement that two ratios are equal.

Exercise:
a) 18% b) 75% c) 12 1/2% d) 3% e) 7%

Exercise:

1) 24/100 = 24%
2) 13/100 = 13%
3) 13/100 = 13%
4) 13/100 = 13%
5) 11/100 = 11%
6) 26/100 = 26%

Exercise:

1) 4/5 = 80/100 = 80%
2) 3/20 = 15/100 = 15%
3) 4/25 = 16/100 = 16%
4) 8/10 = 80/100 = 80%
5) 7/50 = 14/100 = 14%

Exercise:

a) $\frac{7}{8} \times \frac{12\ 1/2}{12\ 1/2} = \frac{87\ 1/2}{100} = 87\ 1/2\%$

b) $\frac{1}{3} \times \frac{33\ 1/3}{33\ 1/3} = \frac{33\ 1/3}{100} = 33\ 1/3\%$

c) $\frac{5}{6} \times \frac{16\ 2/3}{16\ 2/3} = \frac{83\ 1/3}{100} = 83\ 1/3\%$

d) $\frac{2}{9} = \frac{2}{9} \times \frac{11\ 1/9}{11\ 1/9} = \frac{22\ 2/9}{100} = 22\ 2/9\%$

e) $\frac{3}{11} = \frac{3}{11} \times \frac{9\ 1/11}{9\ 1/11} = \frac{27\ 3/11}{100} = 27\ 3/11\%$

Exercise:

1)
N/1000 = 24/100
100N = 24,000
N = 240 students

2)
N/30 = 6/100
100N = 180
N = 1.8 rainy days

3)
N/10 = 40/100
100N = 400
N = 4 games won

4)
$\frac{2}{3} \times \frac{33\ 1/3}{33\ 1/3} = \frac{66\ 2/3}{100}$

$\frac{N}{12} = \frac{66\ 2/3}{100}$

100 N = 800
N = 8 oranges

OBJECTIVE 6-93

Pretest:

1) a) $N/100 = 5/10$
$10N = 500$
$N = 50\%$

b) $N/100 = 100/25$
$25N = 10{,}000$
$N = 400\%$

c) $N/100 = 21/300$
$300N = 2100$
$N = 7\%$

d) $N/100 = 80/500$
$500N = 8000$
$N = 16\%$

Exercise:

1) $N/100 = 3/5$
$5N = 300$
$N = 60\%$ (peanut butter)

$N/100 = 2/5$
$5N = 200$
$N = 40\%$ (molasses)

2) $N/100 = 8/10$
$10N = 800$
$N = 80\%$ (to friends)

$N/100 = 2/10$
$10N = 200$
$N = 20\%$ (for himself)

OBJECTIVE 6-94

Pre test:

a)
$10/100 = 14/N$
$10N = 1400$
$N = 140$

b)
$6/100 = 12/N$
$6N = 1200$
$N = 200$

c)
$N/100 = 5/10$
$10N = 500$
$N = 50$

d)
80/100 = 10/N
80N = 10000
N = 12 1/2

e)
N/100 = 2/20
20N = 200
N = 10

2) Answers will vary.

Exercise:

1)
10/100 = 95/N
10N = 9500
N = 950 students

2)
20/100 = 10/N
20N = 1000
N = 50 comics

3)
25/100 = 6/N
25N = 600
N = 24 or 2 dozen eggs

4)
33 1/3/100 = 50/N
33 1/3N = 5000
N = $1.50

5)
200/100 = 9/N
200N = 900
N = $4.50

OBJECTIVE 6-95

Pretest:

1) 2
2) yes
3) no
4) part of one hundred
5) The whole or 100% of a number.

ANSWER KEY—LEVEL 7

OBJECTIVE 7—1

Pretest:

1-6) Verbal answers.
7) Five hundred eighty-nine.

8) Nine trillion, two hundred seven billion, eight hundred million.
9) Eight hundred forty-three million, two hundred ninety-seven thousand, eight hundred sixty-one.
10) 2,058,767,593

Exercise:
Balance:
Three billion, one hundred seventy-six million, six hundred eighty-nine thousand, nine hundred forty-three.
Deposits:
Thirty-three billion, seven hundred eighty-nine million, nine hundred sixteen thousand, one hundred ninety-eight.
Withdrawals:
Forty-one billion, three hundred seventy-eight million, eight hundred ninety-seven thousand, seven hundred fifty.
Total debt:
Two hundred eighty billion, three hundred fifty million, eight hundred sixty-six thousand, one hundred twenty-nine.
Gold assets:
Twenty-one billion, six hundred ninety-two million, seven hundred six thousand, two hundred thirty-eight.

OBJECTIVES
7-(14, 15)

Pretest:
1) Clock dials, outlines, preface, book no., chapter no. dates, etc.
2) a) 5 b) 50 c) 500 d) 2000 e) 400
f) 90 g) 66 h) 98 i) 1766 j) 3333
3) a) VII b) IV c) XIX d) XXXIV e) XXXVIII
f) LXIII g) LXXVI h) CCXCIX i) CDLIX j) MCMLXX

Exercises:
1) a) 4 b) 60 c) 309 d) 1055 e) 2444 f) 999
2) a) VI b) IX c) XII d) XIX e) XXIX f) LXXVII
g) DCLXVI h) CL i) DCCCIV j) XLIX

OBJECTIVES 7-(18-20)

Pretest:

1. a) 1,2,7,14 b) 1,2,3,4,6,12 c) 1,2,3,6 d) 1,2,3,6,9,18
   e) 1,2,4,8,16,32
2. a) 6 b) 6 c) 12 d) 24 e) 12
3. a) 1 1/6 b) 5/6 c) 5/12 d) 11/24 e) 3/4
4. a) 2 3/4 b) 3 7/12 c) 6 19/24 d) 2 9/20 e) 1 17/30

Exercises:

1. a) 12 b) 8 c) 14 d) 12 e) 6
2. a) 7/12 b) 5/8 c) 3/14
3. a) 7/3 b) 15/4 c) 7/6 d) 41/8 e) 11/5
4. a) 2 7/12 b) 2 3/8 c) 2 5/12

OBJECTIVES 7-(21-23)

Pretest:

1. a) 3 1/4 b) 6 1/2 c) 2 2/3 d) 4 5/6 e) 7 4/5
2. a) 1 2/3 b) 3 3/4 c) 5 2/7 d) 5 7/8 e) 7 1/2
3. a) 4 2/3 b) 1 1/2 c) 1 7/8 d) 2 1/3 e) 4 4/5

Exercise:

1) 9 4/5 2) 24 5/6 3) 3 2/3

Exercise:

1) 1 3/4 2) 9 5/7 3) 5 5/8
4) 4 2/3 5) 17 3/4 6) 13 5/6

OBJECTIVES 7-(24-28)

Pretest:

1) a) 1/6	b) 8/21	c) 1/32
2) a) 3 1/2	b) 4	c) 1
3) a) 1 1/8	b) 1 23/25	c) 1 1/2
4) a) 9	b) 14 1/2	c) 22 3/4
5) a) 13 11/28	b) 5 5/7	c) 11

Exercises:

a) 3/16	b) 2/21	c) 2 4/7	d) 5 1/4	e) 2 1/6
f) 2 3/16	g) 33 1/4	h) 26	i) 8 1/4	j) 45 3/5

OBJECTIVES 7-(29-33)

Pretest:

1) a) 16	b) 21	c) 6
2) a) 3/4	b) 4/5	c) 1 1/2
3) a) 9/16	b) 2 1/9	c) 1/3
4) a) 3 1/3	b) 9 3/16	c) 19 1/4
5) a) 2 19/28	b) 5 1/4	c) 3 1/6

Exercise:

1.	a) 2	b) 4/3	c) 5/2	d) 8/3	e) 5/4
	f) 4/9	g) 3/20	h) 4/7	i) 5/37	j) 8/33
2.	a) 8	b) 2 2/3	c) 1 3/8	d) 16 2/3	e) 17/18

OBJECTIVE 7-49

Pretest:

Item	*Discount*	*Net price*
1. Shoes	$2	$18
2. Sofa	$60	$240
3. Coats	$3	$17
4. Chair	$8	$32
5. Chair	$18	$42

Exercises:

Item	*Discount*	*Net Price*
1. Winter dress	$3	$17
2. Sweater	$1.50	$8.50
3. Radio	$9	$51
4. Shoes	$2.70	$15.30

5. Half-price sale means 50% discount. Amount of discount is $4.75.

OBJECTIVES 7-(51, 52)

Pretest:

1. $648 commission
2. $256 commission
3. 6% is the rate of commission
4. $90 commission

Exercises:

1. $1650 commission
2. $18
3. $300
4. Amount of sale = $700
   He earned $21 commission

Exercises:

1. 5% commission.
2. 4% commission charged by the real estate company.

OBJECTIVE 7-53

Pretest:

Table

1. $7 interest
2. $14 interest
3. $3 interest
4. $96 interest
5. $36 interest
6. $5 interest
7. Owed $72 interest. Amount due = $672

Exercise:

1. Bob pays $2 interest.
2. Sue pays $32 interest.
3. Mary pays $12 interest.
4. Jeff pays $ 39 interest.
5. Tim pays $135 interest.
6. Every 6 months you get $2 interest.
7. Bert gets $5.40 interest. Amount is $185.40.

OBJECTIVE 7-57

Pretest:

Check each student's work for accuracy.

A) 1. The Total should be $212.86
   2. The Total should be $ 58.35

B) 1.	The amount deposited should be	$271.21
	Stub's Total for check No. 1 should be	$271.21
	The Amount this check	$ 61.26
	The Balance carried forward is	$209.95
2.	The Balance brought forward is	$209.95
	Amount deposited is	$ 00.00
	Stub's Total is	$209.95
	Amount this check is	$ 30.55
	Balance carried forward is	$179.40

Exercise:

The total should be $389.25

Bal. Bro't For'd	$ 00.00
Am't Deposited	$389.25
Total	$389.25
Am't this check	15.75
Bal. Car'd For'd	$373.50

OBJECTIVES 7-(60, 61)

Pretest:

1. a) 800 b) 793 c) 360 d) 943
2. a) 08567 b) 86137
3. $ 15.26

Problems:

1. Bill is correct.
2. $19.81
3. 340
4. Bill is incorrect.

OBJECTIVES 7-(62,63)

Pretest:

1) a) 37 b) 211 c) 394 d) 149 e) 256
2) a) 7541 b) 3590
3) 3,700 cubic feet

OBJECTIVE 7-93

Pretest:

1. Factor
2. Base
3. $2^3 \times 3^2 = 8 \times 9 = 72$
4. $3^3 \times 2^2 = 27 \times 4 = 108$
5. $(4^2)(4^2) = 4^4 = 4 \times 4 \times 4 \times 4 = 256$

Exercise:

1. 4 x 4 x 4 = 64
2. 8 x 8 = 64
3. 2 x 2 x 2 x 2 = 16
4. 100 x 100 x 100 x 100 x 100 x 100 x 100 x 100
5. 14 x 14 x 14
6. 25 x 25
7. 2 x 2 x 2 x 2 x 5 x 5
8. 3 x 3 x 3 x 3 x 3
9. 15 x 15 x 2 x 2 x 2 x 2
10. 1 x 5 x 5 x 5 x 5 x 5 x 5 x 5 x 5 x 5 x 5 x 5 x 5 x 5

Exercise:

1) 2^6 2) 3^9 3) 10^6 4) 25^9 5) 1^{14}

OBJECTIVES 7-(100-103)

Pretest:

1) a) hecto- b) deci- c) kilo- d) milli- e) centi-
   f) deca- or deka-
2) The meter

3) a) 10^{-3} b) 10^{0} c) 10^{2} d) 10^{-1} e) 10^{-2}
f) 10^{1} g) 10^{3}

SOLUTION TO WORD PUZZLE

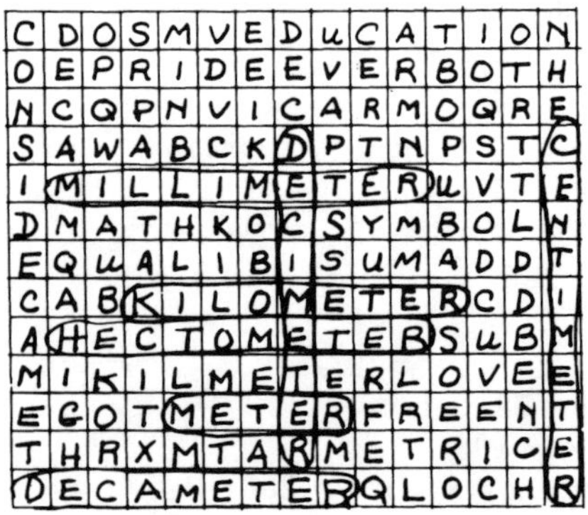

1) 39.37 inches, meter
2) 1/100 meter, centimeter
3) 1000 x a meter, kilometer
4) 10^{-1} meter, decimeter
5) 10 x a meter, decameter
6) 10^{-3} meter, millimeter
7) 100 x a meter, hectometer

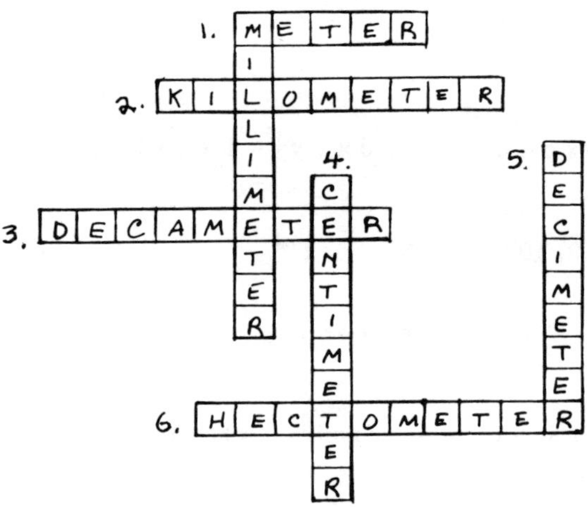

Exercise:

1) 134,000
2) 4359
3) 2345
4) 67,000,000
5) 43,210
6) 8.6754
7) 439,765.4

Exercise:

1) 750 2) .23456 3) 465.029 4) .032 5) .326896
6) 3.6 7) .14362 8) .000231 9) .00000029 10) .11111111

Exercise:

1) 4.5 hm. 24.98 km. .65 hm. (Divide by 100)
2) 1.675 km. 73.6 km 1.9km. (Divide by 10)
3) 1000 mm. = 1 meter
   1000 meters = 1 kilometer
   1 kilometer = 1000 x 1000 = 1,000,000 mm.
   5 km.= 5 x 1,000,000 = 5,000,000 mm.
   .5 Km. = .5 x 1,000,000 = 500,000 mm.
   16 km.= 16 x 1,000,000 = 16,000,000 mm.
4) 1 kilometer = 1,000,000 mm. (See above.)
   $6{,}000{,}000 \div 10^6 = 6$ km.
   $25{,}000 \div 10^6 = .025$ km.
   $250{,}000 \div 10^6 = .25$ km.
5) 1 m. 8 cm = 1.08 m.
   3 cm. 2 mm. = 32 mm.
   5 m. 8 dm. 5 cm. = 585 cm.
   7 cm. 3 mm = 7.3 cm.
   4 m 7 cm. 2 mm. = 4.072 m

ANSWER KEY — LEVEL 8

OBJECTIVE 8-8

Pretest:

1) 19	6) 2
2) 23	7) −1
3) 1	8) −3
4) 1	9) 2
5) 5	10) 48

Exercise:

1) 49	6) −2
2) 4	7) −33
3) 2	8) −204
4) 15	9) −69
5) 10/39	10) 52

Extra for Experts:

1) x	2) $15a^2 - 3a$	3) $3z^2 + 6z$	4) a + 12bc + 6d	5) r + s − 24s
6) x − z	7) 8bd − a − 1	8) 1 /x	9) 1	10) 1

OBJECTIVES 8-19

Pretest:

1) 1
2) a fraction (5/5, 2/2, etc.)
3) Answers will vary.
4) a) 1/2 b) 2/3 c) 1/3 d) 1/3
   e) 1/3 f) 4/5 g) 1/7 h) 1/6
   i) 1/12 j) 8/9

Exercises: Answers will vary.

OBJECTIVE 8-23

Pretest:

1) 2/9
2) 1/6
3) 5/9
4) 4
5) 2/15
6) 5 7/16
7) 1 7/8
8) 2/5
9) 12 3/4

Exercise:

1) 2/7
2) 4/5
2) 3
4) 7/22
5) 1/8
6) 9/25

OBJECTIVE 8-37

Pretest:

1) a) 10^3
   b) 10^5
2) a) 100
   b) 10,000
   c) 1
3) a) (2 x 100) + (3 x 10) + (2 x 1) *OR*
   $(2 \times 10^2) + (3 \times 10^1) + (2 \times 10^0)$
   b) (3 x 10,000) + (4 x 1000) + (7 x 100) + (5 x 10) + (8 x 1) *OR*
   $(3 \times 10^4) + (4 \times 10^3) + (7 \times 10^2) + (5 \times 10^1) + (8 \times 10^0)$
   c) (9 x 10,000,000) + (3 x 1,000,000) + (2 x 100,000)
   + (1 x 10,000) + (4 x 100) + (3 x 10) + (5 x 1)
   OR
   $(9 \times 10^7) + (3 \times 10^6) + (2 \times 10^5) + (1 \times 10^4) + (4 \times 10^2)$
   $+ (3 \times 10^1) + (5 \times 10^0)$
4) a) 37,525
   b) 9, 308,462
   c) 130,650

Exercise:

1) (2 x 1000) + (5 x 100) + (8 x 10) + (3 x 1)
2) (1 x 100,000) + 7 x 1000) + (6 x 100) + (9 x 10) +(3 x 1)
3) (5 x 1,000,000) + (2 x 100,000) + (7 x 1000) + (9 x 100)
   + (3 x 10) + (6 x 1)

4) (7 x 10,000) + (4 x 1000) + (1 x 100) + (2 x 1)
5) (8 x 10,000,000) + (2 x 1,000,000) + (5 x 100,000) + (9 x 10,000) + (3 x 1000) + (4 x 100) + (2 x 10) + (6 x 1)

Exercise:

1) a) 2
   b) 5
   c) 100
   d) no
2) a) 3
   b) 4
   c) 6
   d) 10
3) a) 10^5
   b) 10^8
4) a) 1,000,000
   b) 100,000,000
   c) 1,000,000,000,000,

Exercise:

1) a) $(3 \times 10^6) + (5 \times 10^5) + (6 \times 10^3) + (2 \times 10^2) + (1 \times 10^1) + (7 \times 10^0)$
   b) $(9 \times 10^5) + (4 \times 10^4) + (5 \times 10^3) + (7 \times 10^2) + (2 \times 10^0)$
2) a) $(7 \times 10^4) + (3 \times 10^2) + (4 \times 10^1) + (2 \times 10^0)$
   b) $(9 \times 10^7) + (6 \times 10^6) + (5 \times 10^5) + (3 \times 10^4) + (4 \times 10^2) + (2 \times 10^1) + (1 \times 10^0)$
3) 82, 530,926

OBJECTIVE 8-38

Pretest and Activities: All verbal answers.

OBJECTIVE 8-40

Pretest:

1) .726
2) 16.14
3) 4.005
4) 143.313
5) .0275

Multiply:

1) 26.949117
2) 31.536
3) 10.9989
4) 53.20692408
5) 110.65846
6) 7.434
7) 64.2910576
8) 28.391502
9) 146.4146068
10) 93.7392

OBJECTIVE 8-41

Pretest:

1) 2
2) 900
3) 70,000
4) .6
5) 21
6) 60,000
7) 880
8) 6.03
9) 131,100
10) 10

Exercise:

1) $6.23 \times 10^2 = 623$
2) $.058 \times 10^3 = 58$
3) $29.6 \times 10^1 = 296$
4) $145{,}089 \times 10^0 = 145{,}089$
5) $.07 \times 10^2 = 7$
6) $17.2 \times 10^1 = 172$
7) $.005 \times 10^3 = 5$
8) $.033 \times 10^3 = 33$
9) $.0102 \times 10^4 = 102$
10) $3.0204 \times 10^4 = 30{,}204$

Exercise:

1) .6
2) .06
3) .6
4) 6000
5) 60,000
6) .06

OBJECTIVE 8-66

Pretest:

1) 9.3×10^7
2) 1.57×10^3
3) 1.2156×10^7
4) 1.00025×10^5
5) 1.560015×10^4
6) 1.0015×10^2
7) 1.9145×10^4
8) 2.01007×10^2

Exercise:
1) 1.56×10^{10}
2) 9.875×10^{12}
3) 1.007×10^{12}
4) 1.56817×10^2
5) 3.14875×10^{13}

Exercise:
1) 3.3×10^{-3}
2) 1.98×10^{-1}
3) 7×10^{-4}
4) 1.23456×10^{-1}
5) 5.0005×10^{-4}

OBJECTIVE 8-91

Pretest:
1) b
2) a = 4
2) Answers will vary. (An *equation* is an open number sentence which states that two quantities are equal.)
4) Answers will vary. (A *variable* is a symbol that takes the place of an unknown number.)

Opposites:

−35 −109 + 32 + 1/4 −.25 + 1 1/2 −75

Exercise:

1) 3
2) 6
3) 8½
4) 9
5) 5
6) 11

Challenge:
a = 6

OBJECTIVE 8-114

Pretest:

1) a) one hundredth
   b) one thousandth
   c) one tenth
   d) one thousand
2) . 275 kilograms
3) 60.8 liters
4) 1000
5) 1.1

Exercise:

1) 26.95 (27) Quarts
2) 9.5 Liters

Exercise:

1) 1,832,000 grams
2) .0000035 oz.
3) 20 lbs.

OBJECTIVE 8-118

Pretest:

1) .7	5) 1.969	9) 150	13) 46,000
2) 5.4	6) 3.247	10) 90	14) 9,123,000
3) 8.35	7) 57	11) .008935	
4) .51	8) 90	12) 2.050030	

Exercise:

1) 9000
2) 3,540,000
3) 1.008
4) .0075
5) 24

APPENDIX II

ANSWER KEY

POSTTESTS

LEVEL 6

POSTTESTS

OBJECTIVES 6-(71,72)

Posttest:

1. Opposite
2. zeros
3. a) 100
   b) 4100
   c) 2890
   d) 8000
   e) 1,000,000
   f) 896.05
   g) 4890
   h) 1616
   i) 1,085,000
   j) 246
4. a) 1
   b) 10
   c) 4.68
   d) .1234
   e) .05623
   f) 1000
   g) .1
   h) .001
   i) 1
   j) .00087

OBJECTIVE 6-74

1) Answers will vary.
2) 625 225 5625
3)

64	49	16	9	81	36
784	729	1936	2809	5184	6561
15.21	23.04	2510.01	636.5529		

4)

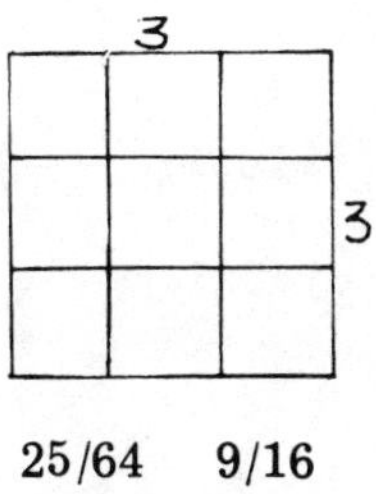

5) 1/9 25/64 9/16 $\frac{9}{8} \times \frac{9}{8} = \frac{81}{64} = 1\frac{17}{64}$

OBJECTIVE 6-79

Posttest:

1) a) 1,2,3,4   b) 1,2,3,4   c) 1,2,3,4   d) 1

2)

332	9000	91,836	
93,111	501		
16,505	12,895		
624	6,186	42,654	
5,888	96,968	327,728	
3,456	9,639	999,981	
2,500	18,320	10,100	180,320

3) Will vary.

OBJECTIVES 6-(80-83)

Posttest:

1. The same as the number of the base.
2. Binary
3. $17 = 10001_2$
   $63 = 111111_2$
   $47 = 101111_2$
   $55 = 110111_2$
   $38 = 100110_2$
   $22 = 10110_2$
4. a) 7
   b) 11
   c) 17
   d) 63
   e) 31

Challenge: 64   128   256   512.

OBJECTIVE 6-84

Posttest:

1. a) 11110_2 b) 1111_2 c) 10110_2 d) 10111_2 e) 1100101_2

2. a) 100_2 b) 10010_2 c) 111_2 d) 1001_2 e) 1

OBJECTIVE 6-85

Posttest:

1. a) 100011_2
   b) 11000_2
   c) 110001_2
   d) 1000110_2
   e) 1010111110_2

2) a) 1010_2 R = 1
   b) 1011_2
   c) 1000_2
   d) 1111_2
   e) 1011_2

OBJECTIVES 6-(86-88)

Posttest:

1. a) one hundred b) one tenth c) ten d) one thousand
   e) one thousandth f) one hundredth
2. a) 10^2 b) 10^{-1} c) 10^1 d) 10^3 e) 10^{-3} f) 10^{-2}
3. a) decimeter
   b) hectometer
   c) kilometer
4. 1/10; 1/100; 1/1000
5. Answers will vary.

OBJECTIVE 6-(91, 92)

Posttest:

1. 25/100 1½/100 19/100 83 1/3/100 100/100
   250/100 37½/100 3/100 103/100
2. 50/100 = 1/2 40/100 = 2/5 15/100 = 3/20 75/100 = 3/4
   80/100 = 4/5 20/100 = 1/5 35/100 = 7/20 85/100 = 17/20
3. A comparison of two numbers by division.
4. A statement that two ratios are equal.
5. Will vary.

6) a) N/100 = 38/100
   100N = 3800
   N = 38 chicken
   100 − 38 = 62 beef.

b) $N/9 = 20/100$
$100N = 180$
$N = \$1.80$ off

c) $N/35 = 10/100$
$100N = 350$
$N = \$3.50$

OBJECTIVE 6-93

Posttest:

1) 100%
2) Percent sign.
3) a) $N/100 = 13/65$
$65N = 1300$
$N = 20\%$

b) $N/100 = 38/114$
$114N = 3800$
$N = 33\ 1/3\%$

c) $N/100 = 16/96$
$96N = 1600$
$N = 16\ 2/3\%$

d) $N/100 = 125/100$
$100/N = 12500$
$N = 125\%$

OBJECTIVE 6-94

Posttest:

1) a) $N = 21$ b) $N = 100$ c) $N = 7$
2) a) $20/100 = 1/5$ b) $25/100 = 1/4$ c) $33\ 1/3\ /\ 100 = 1/3$
d) $85/100 = 17/20$ e) $125/100 = 5/4 = 1\frac{1}{4}$ f) $10/100 = 1/10$
g) $75/100 = 3/4$ h) $50/100 = 1/2$ i) $80/100 = 4/5$

3) a) 10/100 = 2/N
10N = 200
N = 20

b) 50/100 = 25/N
50N = 2500
N = 50

c) 25/100 = 18/N
25N = 1800
N = 72

d) 25/100 = 125/N
25N = 12,500
N = 500

e) 5/100 = 200/N
5N = 20,000
N = 4000

OBJECTIVE 6—95

Posttest:

1) a) 25/100 = N/176
4400 = 100N
44 = N

b) N/1000 = 16/100
100N = 16,000
N = 160

c) N/75 = $33\frac{1}{3}$ / 100
100N = 2500
N = 25

d) 3/4 = 75/100
N/128 = 75/100
100N = 9600
N = 96

e) 3/8 = 37 1/2 / 100
N/72 = 37 1/2/100
100N = 2700
N = 27

2) a) 35/70 = N/100
3500 = 70N
50 = N
Ans. 50%

b) N/100 = 33/165
165N = 3300
N = 20
Ans. 20%

c) N/100 = 8/80
80N = 800
N = 10
Ans. 10%

d) N/100 = 9/12
12N = 900
N = 75

e) N/100 = 72/80
80N = 7200
N = 90

3) a) 25/100 = 16/N
25N = 1600
N = 64

b) 75/100 = 19/N
75N = 1900
N = 25 1/3

c) 100/100 = 100/N
100N = 10,000
N = 100

d) 10/100 = 13/N
10N = 1300
N = 130

e) 15/100 = 45/N
15N = 4500
N = 300

4) 33 1/3 /100 = 1/N
33 1/3 N = 100
N = $3 weekly allowance

5) N/180 = 10/100
100N = 1800
N = 18 days absent

180 − 18 = 162 days present

6) N/100 = 162/180
180N = 16,200
N = 90%

7) N/100 = 7/28
28N = 700
N = 25%

8) 75/100 = 18/N
75N = 1800
N = 24 words on test

9) 75/100 = 6/N
75N = 600
N = 8 close friends

10) 20/100 = N/30
600 = 100N
6 = N mystery stories

POSTTESTS—LEVEL 7

OBJECTIVE

Posttest:

Answers 1-5 verbal.

6) Sixty-eight.
7) Four thousand nine hundred seventy-five.
8) Three million, five hundred eighty-two thousand, nine hundred forty-two.
9) Seven quadrillion, eight hundred trillion, one hundred fifty-two billion, nine million.
10) Pacific Ocean: 63,801,668 square miles.
11) Atlantic Ocean: 31,830, 718 square miles.
12) Difference in area is 31,970,950 or thirty-one million, nine hundred seventy thousand, nine hundred fifty square miles.
13) A googol is 10^{100}. It has 33 periods having only zeros, and the 34th period has a *10* in it.

OBJECTIVES 7-(14, 15)

Posttest:

1. Answers will vary.
2. Answers will vary.
3. a) 7 b) 13 c) 14 d) 39 e) 44 f) 77 g) 441 h) 1965 i) 3599 j) 1976
4. a) XIV b) XVII c) XLII d) LXXXIX e) XCVIII f) CCXII g) DLV h) CMXCIX i) MCMXCV j) MMDCCXCVIII
5. a) XV b) IX c) L d) M e) CC f) XLV

OBJECTIVES 7-(18-20)

Posttest:

	a)	b)	c)	d)	e)
1.	12	24	12	21	15
2.	19/24	31/42	7/8	11/12	23/24
3.	2 19/21	3 7/12	1 11/28	4 7/24	6 4/9

OBJECTIVES 7-(21-23)

Posttest:

1. a) 11 8/9	b) 16 3/11	c) 8 1/5	d) 10 5/8	e) 14 7/8
2. a) 2 1/4	b) 6 13/24	c) 4 3/28	d) 13 17/18	e) 6 3/7
3. a) 11 6/7	b) 9 3/11	c) 25 1/5	d) 6 3/4	e) 4 5/8

OBJECTIVES 7-(24-28)

Posttest:

1. a) 1/20	b) 4/15	c) 9/32	d) 1/36	e) 9/40
2. a) 4	b) 2	c) 9	d) 14	e) 2 4/7
3. a) 7/8	b) 5/6	c) 1 11/40	d) 2 7/12	e) 1 29/45
4. a) 15 3/4	b) 18 2/3	c) 13 1/2	d) 48 6/7	e) 58
5. a) 5 5/8	b) 29 37/40	c) 30 9/32	d) 17 3/16	e) 25 3/8

OBJECTIVES 7-(29-33)

Posttest:

1. a) 5 1/3	b) 17 1/2	c) 98	d) 31 1/4	e) 18
2. a) 7/8	b) 6	c) 1 2/3	d) 16/21	e) 1 3/5
3. a) 1 1/24	b) 33/80	c) 1 3/10	d) 2 8/9	e) 5/32
4. a) 8 1/6	b) 26 2/3	c) 5 15/16	d) 6 3/14	e) 41 3/5
5. a) 2 5/11	b) 3 1/33	c) 1/2	d) 2	e) 1 16/21

OBJECTIVE 7-49

Posttest:

Item	Amount of discount	Net price
1. Ring	\$.60	\$5.40
2. Necklace	\$.55	\$4.95
3. Tennis racket	\$4.80	\$27.20
4. T.V.	\$120	\$360
5. Refrigerator	\$57	
6. Cassette recorder	A-Mart \$30 Bradlers \$24	\$90 \$96

A-Mart is the cheapest. The recorder costs \$6 less than at Bradlers.

OBJECTIVES 7-(51,52)

Posttest:

1. $200
2. $7,500
3. $21
4. Commission is $870. Jones receives $13,630.
5. 7% (commission charged by real-estate company)
6. 6% (Mr. Town's rate of commission)

OBJECTIVE 7-53

Posttest:

1) $9 Simple interest
2) $12.50
3) $104.00
4) $9.60
5) $843.75
6) Mr. Jennings receives $5 every 1/2 year. The amount after 5 years is $250.
7) Mr. Jennings pays $12 interest. Amount = $212.
8) Bank A pays $42; B pays $56; C pays $42.
   B gives the best return.

OBJECTIVE 7-57

Posttest:

1. The total should be $221.19

2.

Bal. Bro't For'd	$ 00.00
Am't deposited	$221.19
Total	$221.19
Am't this check	$ 45.92
Bal. Car'd For'd	$175.27

3.

Bal. Bro't For'd	$175.27
Am't deposited	$ 00.00
Total	$175.27

Am't this check	$ 54.27
Bal. Car'd For'd	$121.00

4. $121.00; $371.00

OBJECTIVES 7-(60, 61)

Posttest:

1. a) 50141 b) 43705
2. a) $33.15 b) The bill is correct
3. Answers will vary.

OBJECTIVES 7-(62, 63)

Posttest:

1) a) 4607 b) 8529
2) a) $ 51.51 b) The bill is correct (check work).
3) Answers will vary.

OBJECTIVE 7-93

Posttest:

1-3. Will vary.

4) 5^9

5) $1^9 = 1$

6) 1 to any power is always 1.

7) a) 32 b) 27 c) 1024
d) 100,000 e) 1 f) 49
g) 54 h) 243 i) 144
j) 180 k) 4608 l) 15,625

OBJECTIVES 7-(100-103)

Posttest:

1. Will vary

2. a) 7 m. 61 cm. = 7.61 m.
   b) 15 cm. 8 mm = 15.8 cm.

3. 1 m. 95 cm. = 1.95 m.

4. $A = \pi r^2$
   $A = 3.14 \times (17)^2$
   $A = 3.14 \times 289$
   $A = 907.46$ sq. m.

5. $A = s^2$
   $A = (5 \text{ mm})^2$
   $A = 25$ sq. mm.

6. $P = 2A + 2B$ or $(2l + 2w)$
   $P = 2(16) + 2(13)$
   $P = 32 + 26$
   $P = 58$ m.

7. 55.82 m.
   −54.27 m.
   1.55 m.

8. 135 m. x 60 = 8100 m. (per min.)
   $8100 \text{ m.} \div 10^3 = 8.1$ km.

9. 39.37
   x 2
   78.74 in. in 2 m.
   1 yard = 36 inches
   78.74 ÷ 36 = 2.19 yds.

10. Will vary.

POSTTESTS — LEVEL 8

OBJECTIVE 8-8

Posttest:

1)	5	6)	2	11)	1/3	16)	1/2
2)	8	7)	5	12)	2	17)	73.3
3)	15	8)	10	13)	1 7/8	18)	1
4)	32	9)	5	14)	4 12/13	19)	14 5/28
5)	18	10)	3	15)	30	20)	54

OBJECTIVE 8-19

Posttest:

1) Answers will vary.

2)

a) 4	h) 15	o) 70
b) 5	i) 25	p) 75
c) 15	j) 28	q) 45
d) 40	k) 35	r) 86
e) 18	l) 26	s) 8
f) 9	m) 27	t) 4
g) 34	n) 34	u) 37 ½

OBJECTIVE 8-23

Posttest:

1) 1 1/7	7) 2 5/36
2) 5/6	8) 44 2/3
3) 1 3/4	9) $.93
4) 2	10) 3750 miles
5) 1 2/3	11) $1.32
6) 1 3/4	

OBJECTIVE 8-37

Posttest:

1) Answers will vary.

2) a) 10^6

b) 7×10^4

c) 3×10^4

3) a) (3 x 10,000) + (4 x 1000) +(9 x 100) + (7 x 10) + (2 x 1)

$(3 \times 10^4) + (4 \times 10^3) + (9 \times 10^2) + (7 \times 10^1) + (2 \times 10^0)$

b) (7 x 1,000,000) + (6 x 100,000) + (3 x 10,000) + (4 x 1000) + (1 x 100) + (2 x 10) + (5 x 1)

OR

$(7 \times 10^6) + (6 \times 10^5) + (3 \times 10^4) + (4 \times 10^3) + (1 \times 10^2) + (2 \times 10^1) + (5 \times 10^0)$

c) (8 x 1,000,000) + (3 x 10,000) + (4 x 1000) + (9 x 100) + (2 x 10) + (1 x 1)

OR

$(8 \times 10^6) + (3 \times 10^4) + (4 \times 10^3) + (9 \times 10^2) + (2 \times 10^1) + (1 \times 10^0)$

4) a) 349
b) 670,981
c) 42,057,209

OBJECTIVE 8-38

Posttest:

1. .0	2. 8		3. 1	4. 5	5. 7	6. .6
	7. .7	8. 2	5	1		1
9. .1	4	8	6		10. .2	5
11. 9	3	.6		12. .2	1	5
13. 8		14. 5	15. .5	5	8	8
	16. .3	2	4	3	9	6

OBJECTIVE 8-40

Posttest:

1) a) 2
b) 1
c) 3
d) 3
e) 4
f) 0

2) a) 2.4
b) 6.24
c) 11.4
d) 630
e) .00481
f) . 00126

3) a) 4.93
b) 700.8
c) .1548516
d) .1068
e) .02295
f) 4.7842
g) 1.68
h) 24.258
i) .886122513
j) .0000048
k) 160,900
l) 136.7774781

OBJECTIVE 8-41

Posttest:

1) *Divisor:* number by which another number is being divided.
   Dividend: the number being divided.
   Quotient: the answer in division.
   Remainder: the number left over when a number cannot be evenly divided.

2) .00005

3) a) 0 b) 3 c) 1 d)10 e) 15

4)

a) 25	d) 148.9	g) 3500	j) 300,000
b) 2.5	e) .05	h) 2090	k) 2487
c) .025	f) 330	i) 2.056	l) 9

5) \$.055 *or* \$.05½ *or* 5½¢

6) 18 minutes

7) a) 1900 b) 57 c) 9330 d) 13 e) 900

OBJECTIVE 8-66

1) 3.0×10^{27}

2) 100,000,000,000,000

3) 1.58×10^{18}

4) 1×10^{100}

5) a) 7.123×10^{12}
   b) 6.0×10^{9}
   c) 1.3×10^{13}
   d) 4.35008011×10^{8}
   e) $1.0000000225 \times 10^{10}$

6) a) 7×10^{-1}
   b) 1.1×10^{-3}
   c) $1.00000000002 \times 10^{1}$
   d) $8.90023675 \times 10^{-1}$
   e) 3.9816×10^{-14}

OBJECTIVE 8-91

Posttest:

1) Answers will vary.
2) a) 3 f) 4
   b) 1 g) –16
   c) 46 h) 5
   d) 1/3 i) 1 1/8
   e) 23 j) .88

OBJECTIVE 8-114

Posttest:

1) 681.8 kilograms
2) 15.85 or 16 liters
3) 11.9 kilograms
4) 4545.45 or 4545.5 kilograms
5) less
6) 18.2 kilograms

OBJECTIVE 8-118

Posttest:

1.) 964,342,000,000
2.) $11,900 (or $11,860)
3.) $146
4.) $.08 (8¢)
5.) a. 9.16 = 9.2
    b. 60.82 = 60.8
    c. 1363.63 = 1363.6
    d. 12.85 = 12.9
    e. 3.28 = 3.3

INDEX